SCI PUBLICATION 186

Design of Steel Framed Buildings without Applied Fire Protection

C G Bailey BEng, PhD
G M Newman BSc(Eng), MIStructE, AIFireE
W I Simms BEng, D Phil

Published by:

The Steel Construction Institute
Silwood Park
Ascot
Berkshire SL5 7QN

Tel: 01344 623345
Fax: 01344 622944

© 1999 The Steel Construction Institute

Apart from any fair dealing for the purposes of research or private study or criticism or review, as permitted under the Copyright Designs and Patents Act, 1988, this publication may not be reproduced, stored or transmitted, in any form or by any means, without the prior permission in writing of the publishers, or in the case of reprographic reproduction only in accordance with the terms of the licences issued by the UK Copyright Licensing Agency, or in accordance with the terms of licences issued by the appropriate Reproduction Rights Organisation outside the UK.

Enquiries concerning reproduction outside the terms stated here should be sent to the publishers, The Steel Construction Institute, at the address given on the title page.

Although care has been taken to ensure, to the best of our knowledge, that all data and information contained herein are accurate to the extent that they relate to either matters of fact or accepted practice or matters of opinion at the time of publication, The Steel Construction Institute, the authors and the reviewers assume no responsibility for any errors in or misinterpretations of such data and/or information or any loss or damage arising from or related to their use.

Publications supplied to the Members of the Institute at a discount are not for resale by them.

Publication Number: SCI-P-186

ISBN 1 85942 062 1

British Library Cataloguing-in-Publication Data.
A catalogue record for this book is available from the British Library.

FOREWORD

This publication is intended to assist in the scheme design of buildings, and to show how steel structures can be designed easily to achieve up to 60 minutes fire resistance, without additional protection. It should be of particular interest to Structural Engineers, Architects and Contractors (particularly in design and build structures).

There are now many types of steel beam and column which inherently have up to 60 minutes fire resistance without applied protection. These include various types of slim floor beam, partially encased beams and columns, and concrete filled columns.

The authors of this publication are Dr Colin Bailey of BRE (formerly of SCI), Mr Gerald Newman of The Steel Construction Institute and Dr Ian Simms also of The Steel Construction Institute. The development of this publication was funded by The Department of Environment, Transport and the Regions (DETR) and British Steel Section, Plates & Commercial Steels under the Partners in Technology initiative.

During the preparation of this publication, valuable comments were received from a number of people and SCI would like to thank the following for their contributions:

Dr Mark Lawson	The Steel Construction Institute
Mr Jef Robinson	British Steel (SP&CS)
Mr John Dowling	British Steel (SP&CS)
Mr John Rushton	Peter Brett Associates
Mr Robert Brickwood	Peter Brett Associates
Mr Michael Green	Buro Happold - FEDRA

The Department of the Environment, Transport and the Regions partly sponsored the development phase of this project.

CONTENTS

                                                Page No.

		Page No.
	SUMMARY	vii
1	INTRODUCTION	1
2	REVIEW OF CONSTRUCTION SYSTEMS	2
3	STRUCTURAL FIRE ENGINEERING	3
3.1	Fire resistance	3
3.2	Section factors	3
3.3	Design codes	4
4	BEAM DESIGN	6
4.1	Non-composite unprotected beams supporting concrete floors	6
4.2	Composite beams	7
4.3	Slim floor beams	8
4.4	Shelf angle beams	10
4.5	Partially encased beams	11
4.6	Enhancement of fire resistance of beams using beam-to-column connections	12
5	COLUMN DESIGN	15
5.1	Effective length of columns in fire	15
5.2	Unprotected columns	16
5.3	Blocked-in columns	17
5.4	Partially encased columns	18
5.5	Concrete filled hollow section columns	19
5.6	Brick or block encasement	21
6	FLOOR SLAB DESIGN	22
6.1	Shallow composite floor slabs with steel decking	22
6.2	Deep composite floor slabs	23
6.3	Precast concrete slabs	24
7	STEEL MEMBERS SUPPORTING COMPARTMENT WALLS	25
7.1	Thermal bridging	25
7.2	Effect of structural deformation	25
8	OVERALL FRAME STABILITY	28
8.1	Braced frames	28
8.2	Sway frames	28

9	EXTERNAL STEELWORK	29
10	CONNECTION DETAILS	30
	10.1 General principles	30
	10.2 Connection of internal *Slimflor* beams to columns	30
	10.3 Connection of RHS *Slimflor* edge beam to columns	33
	10.4 Shelf angle floor beam	34
	10.5 Partially encased beams	35
11	DESIGN EXAMPLES	37
	11.1 Beam design	38
	11.2 Column design	44
	11.3 Floor slab design	47
12	REFERENCES	55
APPENDIX A:	BS 5950-8: Limiting temperature method	59
APPENDIX B:	The moment capacity method	60
APPENDIX C:	Design of 'open' car parks in fire conditions.	61
APPENDIX D:	Design Data Sheets	62

SUMMARY

This publication presents the state-of-the-art methods that can be used to design steel framed structures without the need for passive fire protection materials for up to 60 minutes fire resistance. It covers the design of the following types of beam and column and includes design examples:

    Non-composite downstand beams
    Composite downstand beams
    Slim floor beams
    Shelf angle beams
    Partially encased beams

    Unprotected columns
    Blocked-in columns
    Partially encased unreinforced and reinforced columns
    Concrete filled hollow section columns
    Brick or block encasement

Design Data Sheets for most of the design options are presented which facilitate design by providing tabulated information. In some instances more detail may be required, and the reader is directed towards additional published data in the relevant sections.

Le dimensionnement des immeubles à ossature en acier sans protection incendie

Résumé

Cette publication présente l'état de la question concernant les méthodes qui peuvent être utilisées pour dimensionner les ossatures en acier sans avoir recours à des matériaux de protection passive contre l'incendie et permettant d'atteindre une résistance à l'incendie jusqu'à 60 minutes. Elle couvre le dimensionnement des éléments structuraux suivants et donne des exemples pratiques:

    Poutres métalliques sous dalles
    Poutres composites sous dalles
    Poutres pour planchers minces
    Poutres supports en cornières
    Poutres partiellement encaissonnées

    Poteaux non protégés
    Poteaux entourés de béton
    Poteaux renforcés ou non renforcés partiellement encaissonnés
    Poteaux en profils creux remplis de béton
    Poteaux entourés de briques ou de blocs

Des formulaires de dimensionnement sont fournis pour les principales variantes. Ils facilitent le dimensionnement en fournissant des informations à l'aide de tables. Dans certains cas, plus de détail peut être nécessaire et le lecteur est guidé vers d'autres publications fournissant ces informations.

Die Berechnung von Stahlkonstruktionen ohne Brandschutz

Zusammenfassung

Diese Publikation stellt die aktuellen Verfahren zur Berechnung von Stahlbauten vor, unter Verzicht auf passiven Brandschutz bis zu einer Feuerwiderstandsdauer von 60 Minuten. Sie behandelt die Berechnung folgender Träger- und Stützentypen, mit Berechnungsbeispielen:

Unterzüge ohne Verbund
Unterzüge als Verbundträger
Flachdecken-Träger
"Shelf angle" - Träger (I-Träger mit am Steg angebrachten Winkeln)
Teilweise ausbetonierte Träger

Ungeschützte Stützen
Stützen mit ausgemauerten Kammern
Teilweise ausbetonierte (mit oder ohne Bewehrung) Stützen
Ausbetonierte Hohlprofil-Stützen
Ziegel- oder Leichtbeton-Ausmauerung

Datenblätter für die meisten Berechnungsmöglichkeiten werden vorgestellt, die die Berechnung durch tabellierte Informationen erleichtern. In manchen Fällen können genauere Angaben erforderlich werden; der Leser wird in den entsprechenden Abschnitten auf zusätzliche, veröffentlichte Information verwiesen.

Proyecto de edificios aporticados de acero sin protección frente al fuego

Resumen:

Esta publicación presenta métodos actualizados que pueden utilizarse para proyectar estructuras aporticadas de acero sin necesidad de utilizar materiales de protección pasiva frente al fuego y con resistencias de hasta 60 minutos.

Abarca el proyecto de los siguientes tipos de vigas y columnas e incluye ejemplos de proyecto:

Vigas bajo piso no mixtas
Vigas bajo piso mixtas
Vigas Slim floor
Vigas angulares
Vigas parcialmente hormigonadas

Columnas no protegidas
Columnas bloqueadas
Columnas parcialmente protegidas armadas y sin armar
Columnas de sección hueca rellenas de hormigón
Protecciones de ladrillos o bloques

Se presentan hojas con datos para proyecto para la mayoría de las alternativas lo que facilita el proyecto al suministrar información tabulada. En algunos casos pueden necesitarse más detalles y el lector es dirigido hacia publicaciones con datos adicionales en cada sección.

Progettazione di edifici intelaiati in acciaio privi di protezione al fuoco

Sommario

Questa pubblicazione presenta uno stato dell'arte sui metodi che possono essere utilizzati nella progettazione di edifici intelaiati in acciaio con resistenza al fuoco di 60 minuti senza protezione passiva dei materiali. Nella pubblicazione sono trattati diversi argomenti, per i quali sono anche incluse applicazioni progettuali. In dettaglio viene fatto riferimento a:

Travi non composte
Travi composte
Travi in spessore di solaio
Travi con angolari dormienti
Travi parzialmente annegate nel calcestruzzo

Colonne non protette
Elementi verticali di vincolo
Colonne parzialmente annegate nel calcestruzzo con e senza armatura
Colonne in acciaio con sezione tubolare riempita di calcestruzzo
Vincoli di incastro nella muratura o in blocchi in conglomerato

Tabelle progettuali per le più comuni soluzioni progettuali sono presentate al fine di agevolare la progettazione mediante sintetiche informazioni tabulate. In alcuni casi pratici sono necessari maggiori dettagli e pertanto il lettore viene direttamente indirizzato a quei dati addizionali pubblicati negli specifici paragrafi di questa pubblicazione.

Dimensionering av byggnader med stålstomme utan extra brandskydd

Sammanfattning

Denna publikation beskriver de olika metoder på vilket stålkonstruktioner kan dimensioneras för att klara 60 minuters brandmotstånd utan något passivt brandskydd, så som brandskyddsfärg eller inklädnad. I publikationen beskrivs följande typer av balkar och pelare med exempel.

Icke brännbara underliggande balk
Underliggande samverkansbalk
Inbyggd samverkansbalk
Hyllbalk
Delvis inbyggda balkar

Oskyddade pelare
Pelare med murstenar mellan flänsarna
Delvis inbyggda oarmerade och armerade pelare
Pelare av betongfyllda hålrör
Inmurade pelare

Dimensioneringsdata är för de flesta fall beskrivna i tabeller vilket underlättar dimensionering. I några fall kan dock mer data krävas, varför läsaren måste hänvisas till mer utförlig information som finns publicerad i aktuellt avsnitt.

1 INTRODUCTION

Fire safety is an important consideration in the design of multi-storey buildings and this publication describes ways of achieving the necessary levels of safety without the use of applied fire protection. According to the Regulations, the elements of structure of most moderate sized offices require 60 minutes fire resistance when tested in a standard fire resistance test, whilst those in two storey offices normally require only 30 minutes fire resistance. The fire resistance requirements for buildings in the UK are specified in *The Building Regulations*[1]. The requirements vary, up to 240 minutes, depending on the building height, size and use.

In most modern steel framed buildings, the members would typically be protected from the effects of fire by spray or board protection, or intumescent coatings. These methods have proven more economical than the more traditional approach of encasing the steel in concrete or masonry.

The required amount of protection to steel frames is dependent on the type of member, its shape and the fire resistance to be achieved. However, *The Building Regulations* do not specify that a structure must have fire protection, only that it must have fire resistance. This is an important distinction, as many types of steel members can economically achieve up to 60 minutes fire resistance without requiring applied fire protection.

At the design stage, most designers do not consider the method of protecting steel structures from the effects of fire. It is common practice for fire protection to be regarded as a finish, consequently, the normal sequence is to design, construct and then fire protect. However, through the influence of BS 5950-8:1990[2], advances have been made, which offer an alternative approach. This involves the use of fire engineering design which is a more rational rather than prescriptive approach. Fire engineering design methods may impose some limitations, but have the advantage of largely eliminating the fire protection trade from site, which can lead to significant economies and take the fire protection activity off the critical path. As approximately 80% of steel framed buildings require 60 minutes or less fire resistance, the scope for application of these methods in practice is appreciable.

In this publication, methods of designing and constructing steel framed buildings for 30 and 60 minutes fire resistance without applied fire protection are described. In most instances the structural configurations presented in this guide are covered by design codes of practice and design guides published by The Steel Construction Institute or other organisations. Only a summary of the design procedures are thus included, and the reader is directed towards more specialised documents for detailed guidance.

2 REVIEW OF CONSTRUCTION SYSTEMS

The fire resistance periods that can economically be obtained from the structural systems discussed in this publication are summarised in Table 2.1. In some cases longer periods can be attained but this would typically result in larger sections, which may be more expensive than using applied fire protection.

It is worth noting that in England and Wales the installation of sprinklers will often result in the requirements for fire resistance being reduced by 30 minutes. Thus some 90 minutes requirements are reduced to 60 minutes, and some 60 minutes requirements are reduced to 30 minutes if a sprinkler systems is installed.

Detailed information on the use of the types of construction illustrated in Table 2.1 are given in Section 4 for beams, and Section 5 for columns.

Table 2.1 *Fire resistances that can economically be obtained for various structural forms for combinations of beams and columns*

Beam type:	Column type: unprotected column	blocked-in column	partially encased unreinforced	partially encased reinforced	concrete filled hollow sections	protected column
unprotected beam	15	15	15	15	15	15
slim floor systems	15	30	60	60	60	60
shelf angle floor	15	30	60	60	60	60
partially encased	15	30	60	>60	>60	>60
protected beam	15	30	60	>60	>60	>60

3 STRUCTURAL FIRE ENGINEERING

In this Section, the design techniques used to determine the fire resistance of various forms of structural elements are briefly described.

3.1 Fire resistance

Fire resistance is defined in terms of performance in a fire resistance test. All structural systems and fire protection materials will normally have been tested in accordance with the British Standard, BS 476-20:1987 and BS 476-21:1987[3], or an equivalent international standard. In a fire resistance test, elements are evaluated in terms of three criteria: insulation, integrity and loadbearing capacity.

'Insulation', is the ability of an element to resist conducted heat from a fire. 'Integrity', is the ability of an element to resist the passage of flame and hot gases, and load bearing capacity is the ability to carry the applied loading. Floors and walls are considered to be 'separating elements', and are required to meet all three criteria. Beams and columns are essentially line elements and are only required to meet the load bearing criterion unless they are built into a compartment wall or floor and may act as a "hot" bridge (refer to Section 7).

During fire tests, extensive information can be obtained on the thermal and structural response of the structural system. This information can lead to the development of new techniques and design methods.

It is common in Europe to refer to fire resistance in terms of 'REI' where 'R' denotes 'loadbearing capacity', 'E' denotes 'integrity' and 'I' denotes 'insulation'. Thus the requirement for a floor might be written as 'REI 60' meaning that the floor has to meet all three criteria for 60 minutes. In this publication, tabulated data taken from Eurocodes is expressed in terms of R30 or R60 meaning that the data refers to only loadbearing capacity of either 30 or 60 minutes.

3.2 Section factors

The rate of increase in temperature of a linear steel member depends on the ratio of the exposed surface area to the volume of the member per unit length, A_m/V. This ratio is invariably expressed in units of m^{-1} and is known as the 'section factor'. In the UK the section factor is typically represented as the ratio of the heated perimeter to the cross-sectional area, H_p/A. However, both relationships give the same value and the European terminology, A_m/V, will become the standard relationship used in the future.

Members with low section factors will heat up more slowly than members with high section factors, as shown diagrammatically in Figure 3.1.

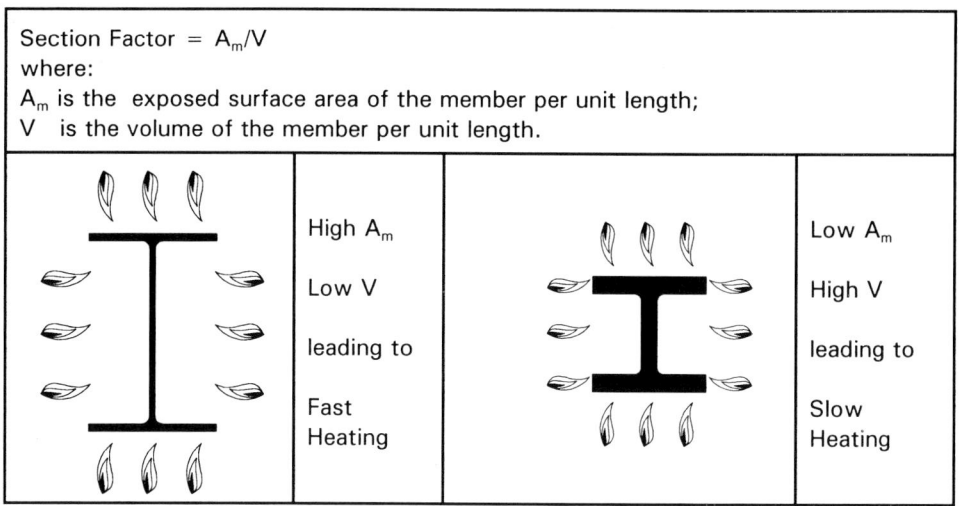

Figure 3.1 *The effect of Section Factor*

3.3 Design codes

BS 5950-8[2], *Code of Practice for fire resistant design* was published in 1990. This was one of the first structural fire design codes in the world, and it contains the general principles for fire design and specific design information for some common structural configurations. During 1999, EC3-1-2[4], which addresses the fire resistant design of steel structures, and EC4-1-2[5], which addresses the fire resistant design of steel and concrete composite structures, were published in the UK as pre-standards or ENVs, together with their National Application Documents (NADs).

All these codes are based on the concept of limit state design. The occurrence of a fire is considered to be an accidental limit state, for which the use of reduced partial load and material safety factors are valid. Therefore, the load intensity is reduced in fire conditions.

3.3.1 Design loads in fire

Statistically, it is accepted that at the occurrence and for the duration of a fire, a member will be subjected to loads less than those used for ultimate design. Therefore, the fire resistant design codes use reduced partial safety load factors, depending on the characteristic of the load. The partial load factors used in BS 5950-8[2] and EC1-1-1 and EC1-2-2[6] are shown in Table 3.1. The major difference in the two codes is in the partial safety factors used for variable imposed loads. EC1 specifies a value of 0.5 for domestic, residential and office buildings, which is increased to 0.7 for congregational and shopping type buildings, and 0.9 for variable loads in storage areas. In comparison, BS 5950-8 specifies a partial safety factor of 0.8 for all variable imposed loads, which increases to 1.0 in escape stairs and lobby areas. The future UK NAD will specify a partial safety factor of 0.7 which increases to 0.9 in escape stairs, lobbies and storage areas.

3.3.2 Design material strengths in fire

BS 5950-8[2] and the Eurocodes[4][5] specify partial factors for steel and concrete for use at the fire limit state. For steel, the value in all codes is 1.0, but for concrete a value of 1.3 is specified in BS 5950-8[2] and a value of 1.0 is specified in EC4-1-2[5]. Based on observed behaviour in fire tests, BS 5950-8 is conservative and the EC4 value is more realistic. The UK NAD for EC4-1-2 allows the EC4 value of 1.0 to be used. The more recent design guides produced by SCI, and computer software for slim floor systems, use a partial factor of 1.0 for all materials at the fire limit state.

Table 3.1 *Partial load factors from BS 5950-8 and EC1-1-1*

Action	BS 5950-8	EC1-1-1
Self weight and dead loads	1.0	1.0
Imposed loads - permanent	1.0	1.0
Imposed loads - variable	0.8	0.5 - 0.9
Wind loads (in combination with imposed loads)	0.33 zero for buildings <8m high	0 or 0.5

3.3.3 Load ratio

The Limiting Temperature method presented in BS 5950-8, is very simple to use, and is mainly based on test data. It uses the concept of load ratio as a non-dimensional measure of the load resisted by a member. It is defined as:

$$\text{Load ratio} = \frac{\text{Load or moment at the fire limit state}}{\text{Member Resistance at 20°C}}$$

After the load ratio is calculated, the limiting temperatures (or maximum allowable temperatures) can be obtained. The use of this method is explained in Appendix A.

In the Eurocodes the load ratio term is expressed as load level or load intensity.

3.3.4 Deflection of elements in fire

At the fire limit state, it is generally accepted that deflections can be very large compared with the normal design limits. For example, a beam may be designed normally for an imposed load deflection of span/360, but in a fire resistance test, a deflection of span/30 may be exceeded. For beams designed to achieve a particular fire resistance, the actual deflections in a real fire will almost certainly be less than those observed in tests. This reduction is caused by complex factors such as restraint from the connections, and by interaction (such as membrane action) with the floor slab. Column deflections are always relatively small and, provided they do not buckle, mid-height displacements are unlikely to exceed height/100 in fire conditions.

It may be necessary to use more severe deflection limits for beams supporting compartment walls or above to ensure the maintenance of the compartmentation function in fire. This may be allowed for in design by imposing lower limits on the temperatures which the elements are allowed to reach, which is intended to reduce deflections of these beams in fire. This typically results in these members, above or below supporting compartment walls, requiring some form of additional fire protection. This is further discussed in Section 7.

4 BEAM DESIGN

This Section presents the methods that can be used to design beams for a fire resistances of 15, 30 and 60 minutes. The four types of unprotected and partially protected beams mentioned in Section 2 are covered; non-composite and composite unprotected beams are discussed separately. Enhancement of fire resistance by utilising the moment capacity of beam to column connections is also discussed.

For each type of beam, the design methods selected are those considered to be the most appropriate. If more than one method is given then the BS 5950-8 methods are given first. If no reference is made to a particular standard, that type of beam is not included in the standard.

In many cases preliminary design information is given in a Design Data Sheet (Appendix D). These are provided to assist in initial design and designers would be expected to refer to more comprehensive information for final design.

4.1 Non-composite unprotected beams supporting concrete floors

Most unprotected Universal Beams (UB) supporting a concrete floor would achieve at least 10 to 20 minutes fire resistance in a standard fire test. Unprotected steel beams can achieve 30 or even 60 minutes fire resistance depending on the beam size and the utilisation of the ultimate limit state capacity This may effectively involve 'over-designing' the section at the ultimate limit state, but this over-design may not be economic, particularly for smaller member sizes.

15 minutes fire resistance

In some car park structures, the steel frame is only required to have 15 minutes fire resistance. Most Universal Beam sections will achieve at least this resistance. For further details see Appendix C.

30 minutes fire resistance

According to BS 5950-8[2], there are three methods for achieving adequate fire resistance for these beams:

1. Ensure that the load ratio is less than or equal to 0.6, and that the section factor does not exceed 90 m^{-1}. (The latter requirement will result in a minimum section size of 610×305×179UB).

2. The Limiting Temperature method in BS 5950-8. This method determines, the maximum temperature at which a beam, for a given load ratio, will sustain the fire design load and compares it with the maximum temperature that the section would reach in the given time. The method is very simple to use and is largely based on test data. For further guidance, see Appendix A.

3. The Moment Capacity method in BS 5950-8. This method is based on plastic design and is therefore only applicable for simply-supported beams which satisfy the requirements for a plastic or compact section as defined in BS 5950-1[7]. To use the Moment Capacity method, the temperature

distribution through the cross-section needs to be known for a given fire resistance period. This is obtained from test data or from thermal models, and is therefore not readily available to most design engineers. For guidance on the use of this method, see Appendix B.

Alternatively, the design method presented in EC3-1-2 can be used. This involves calculating the critical temperature (similar to the Limiting Temperature used in BS 5950-8), based on the applied load at the time of the fire. This Critical Temperature is compared with the maximum steel temperature reached for a given time duration; the maximum temperature is calculated using a simple differential equation. Realistically, this method requires use of a simple computer program or spreadsheet.

60 minutes fire resistance

Unprotected I beams, used as isolated members or supporting concrete slabs, cannot achieve 60 minutes fire resistance without some form of applied fire protection.

4.2 Composite beams

Often the design of composite beams is governed by the serviceability limit state. Therefore, composite beams will often carry a smaller proportion of their ultimate capacity than similar non-composite beams and this reserve of strength can be utilised in fire.

Theoretical studies have shown that the degree of shear connection has some effect on the performance in fire. Beams with low degrees of shear connection should perform better than beams with high degrees of shear connection. However, the effect is small and is normally ignored in assessing the fire resistance of composite beams.

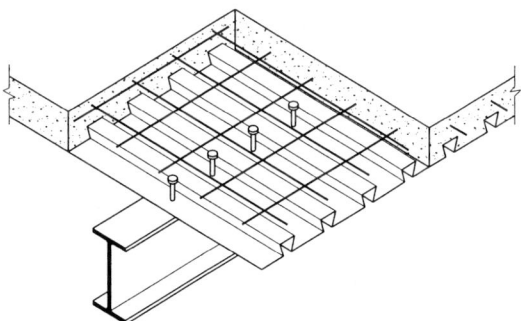

Figure 4.1 *Typical composite beam*

15 minutes fire resistance

In some car park structures, the steel frame is only required to have 15 minutes fire resistance. Most Universal Beam sections used as composite beams will, achieve at least this resistance. For further details see Appendix C.

30 minutes fire resistance

The three methods, of BS 5950-8 described above for non-composite beams, are equally applicable to composite beams.

Alternately, the two design methods presented in EC4-1-2 could be used. The first method presented in EC4-1-2 is similar to, but more conservative than the Limiting Temperature method of BS 5950-8. The second method involves calculating the temperature distribution through the composite beam by splitting the section into elements and using a simple differential equation to calculate the temperature of each element for a given time. This will realistically require a simple computer program or spreadsheet. Based on this temperature distribution the moment resistance can be calculated in accordance with Annex D of EC4-1-2. This method is very similar to the Moment Capacity method in BS 5950-8.

60 minutes fire resistance

Unprotected I beams, when used as part of a composite beam, cannot economically achieve 60 minutes fire resistance and therefore some form of applied fire protection should be used.

4.3 Slim floor beams

Slim floor construction is a recently developed system in which the beam is contained within the depth of the in-situ or precast concrete floor rather than supporting the floor on its top flange. Consequently, all of the steel section, except for the bottom flange or plate is insulated from the fire by the surrounding concrete. In most applications, slim floor beams can achieve 60 minutes fire resistance without applied fire protection.

In the UK, the development of slim floor construction has led to the following forms of construction:

1. A system marketed under the trademark *Slimflor*[8][9]. This consists of a Universal Column section with a plate welded to its bottom flange (designated SFB). The plate is the only exposed part of the beam. The floor can be either precast concrete slabs or deep steel decking which acts compositely with an in-situ concrete topping (Figure 4.2). If hollow core precast concrete slabs are used, then the detailing requirements discussed in Section 6.2 should be followed.

   The design methods for the two forms of *Slimflor* are described in SCI publications *Slim floor design and construction*[8] and *Design of Slimflor fabricated beams using deep composite decking*[9] respectively.

2. A system marketed under the trademark *Slimdek*[10]. This consists of a patented rolled asymmetric steel beam (designated ASB), with the bottom flange wider than the top. A raised rib pattern is formed on the top flange of the beam. The ASB section does not require welding of an additional plate and the section properties have been optimised for design at the ultimate, serviceability and fire limit states. This type of beam (Figure 4.3) has been developed to be used with a deep composite slab using the SD225 deck, supplied by Precision Metal Forming Limited. It has been demonstrated that composite action between the steel and the concrete encasement is achieved for both normal design and in fire. This composite action is enhanced by the

raised rib pattern on the top flange of the beam.

The design of the *Slimdek* system is described in SCI publication *Design of asymmetric Slimflor beams using deep composite decking*[10].

3. Rectangular hollow section *Slimflor* beams[11] (designated RH SFB) comprise an RHS with a plate welded to its lower side. These beams are specifically designed as an edge beam and utilise the good torsional properties of hollow sections to resist the out of balance loads (Figure 4.4).

    The design of *Slimflor* edge beams is described in SCI publication *Design of RHS Slimflor edge beams*[11].

Slimflor and *ASB* sections may be constructed with service holes passing through their webs when used with deep steel decking. The installation of service holes has a detrimental effect on the fire resistance of the beam because it removes part of the web which is fully effective in fire. This typically results in the need for fire protection to be applied to the bottom flange or plate for 60 minutes fire resistance.

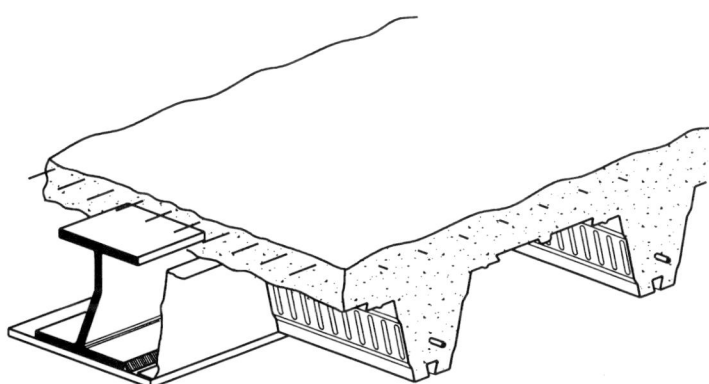

Figure 4.2 *Cut-away view of a Slimflor beam with deep decking*

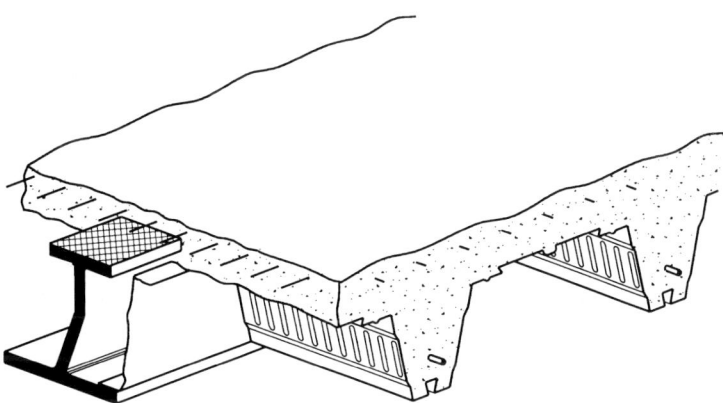

Figure 4.3 *Cut-away view of the Slimdek system using an ASB section*

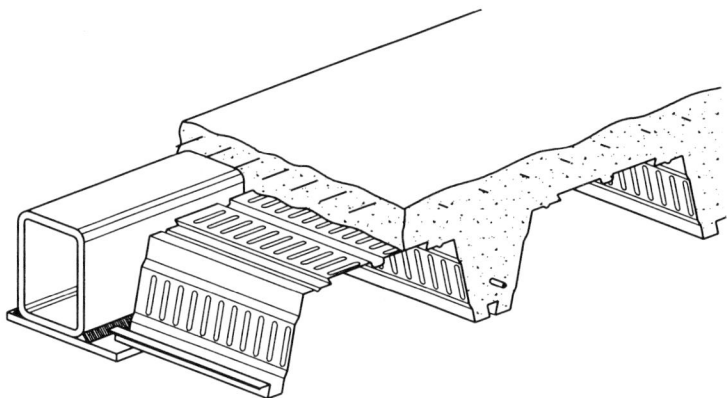

Figure 4.4 *Cut-away view of the RHS Slimflor edge beam*

30 minutes fire resistance

All slim floor beams will achieve 30 minutes fire resistance. For the fabricated sections the plate thickness should be at least 12 mm. Service holes can also be placed in the web of all beams.

60 minutes fire resistance

Two design methods may be used:

1. For beams without openings, the design tables in the relevant SCI publications[8,9,10,11] can be used. Extracts of the tabulated data from these publications are given in Design Data Sheet 1 for *Slimflor*, Data Sheet 2 for *ASB* and Data Sheet 3 for RHS edge beams.

2. SCI design software could be used. For beams with openings, additional fire protection will generally be required to the bottom plate or flange.

In all cases, beams containing service holes cannot normally achieve 60 minutes fire resistance without applied protection to the bottom flange or plate.

4.4 Shelf angle beams

A shelf angle beam consists of a Universal Beam section with angles welded or bolted to the webs to support the floor slab. This type of beam is often used in combination with precast floor units as a means of reducing the overall structural depth. Since the floor units sit on the angles, the top part of the beam is shielded from the fire. For fire resistant design, the angles are positioned with their legs pointing upwards so that they are also shielded from the fire, as shown in Figure 4.5.

Fire resistances of up to 60 minutes may be obtained, although the depth of floor slab required for 60 minutes may result in this form of construction becoming uneconomic. This is because the moment resistance of the beam in fire depends mainly on the portion of the beam which is shielded from fire. For example, for a 406 mm deep beam, the angles would have to be placed at least 210 mm below the top of the beam to achieve a practical load ratio.

The precast concrete units should follow the detailing requirements discussed in Section 6.3.

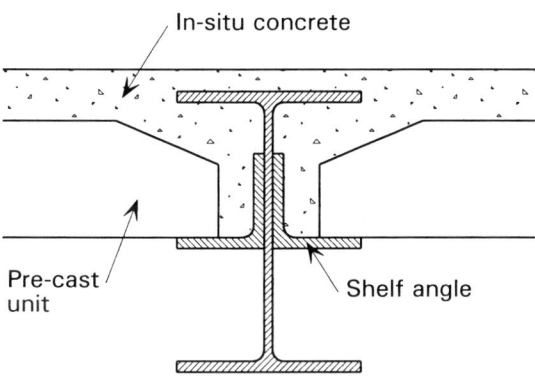

Figure 4.5 *Arrangement of shelf angle beam*

30 minutes fire resistance

Two design methods may be used:

1. Use the design method in BS 5950-8, Appendix E, with temperatures relating to 30 minutes fire resistance. This is based on the moment capacity method discussed in Appendix B of this publication.

2. Use design tables in SCI publication *The fire resistance of shelf angle floor beams to BS 5950: Part 8*[12], which are a direct application of the BS 5950-8 design method. Limited design tables for grade S275 and S355 steel for 30 minutes fire resistance are given in Design Data Sheet 4, of this publication. These give the 'cold' moment capacity and the required position of the shelf angle for a given load ratio and section size. Also presented in the SCI publication 126 are safe load tables for transverse bending of the angles, and tables of longitudinal forces between the beam and the angles.

60 minutes fire resistance

The design methods presented for 30 minutes are also applicable to 60 minutes fire resistance.

4.5 Partially encased beams

A partially encased beam supporting a concrete floor is a form of construction commonly used in continental Europe. The space between the flanges and web is filled with concrete so that the section is rectangular, with the outer surface of the bottom flange exposed. Bar reinforcement may be placed within the concrete, and held in position by shear links (Figure 4.6). The concrete is normally placed before erection of the beams as it is difficult to concrete the sections *in situ*. Local areas for the connections are left exposed and then protected after the frame is erected.

The addition of longitudinal reinforcing bars in the section enhances the load carrying capacity of the member at both the ultimate and fire limit state. However, the use of concrete encasement increases the dead weight of the structure, and may slow down the construction operation.

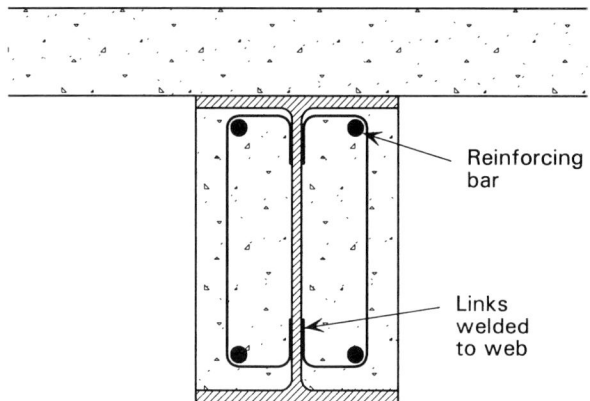

Figure 4.6 *Partially encased composite beam*

30 minutes fire resistance

The design of this type of beam is included in EC4-1-2[5] but is not covered by BS 5950-8. EC4-1-2 gives two methods.

1. For composite beams a tabular method may be used. This is simple to use and is based on load level, section size and percentage of additional reinforcement. Details are given in Design Data Sheet 5. For 30 minutes fire resistance, no additional reinforcing bars are required.

2. A more comprehensive approach is described Annex E of EC4-1-2. This method is based on the moment capacity method described in Appendix B, but is semi-empirical and is based on various tables and formulae. It applies to both composite and non-composite beams.

60 minutes fire resistance

The design methods, presented for 30 minutes are also applicable to 60 minutes fire resistance. Again, often no reinforcing bars are required. If reinforcing bars are used, EC4-1-1[13] states that the links should be welded to the web of the beam or passed through the beam.

4.6 Enhancement of fire resistance of beams using beam-to-column connections

Most multi-storey steel frames in the UK are designed with simple shear resisting connections, and the lateral applied forces are resisted by vertical bracing or shear walls. Under normal loading conditions, moments will be transferred through any steel-to-steel connection. If the connection is 'simple'(as defined in Reference [14]) then the transferred moment is assumed to be small enough to be ignored in the design of the beam. However, the moment transfer which is ignored in normal design offers a reserve that can be utilised in fire. This can provide a significant beneficial effect to beams in fire, as at the fire limit state, high displacements are permitted which would give rise to a noticeable moment transfer through even the most 'flexible' connection. However, the supporting column must be able to support this moment which will require consideration at external members. General guidance is provided in Section 4.6.3.

Because of the possibility of local buckling occurring in beams close to the supports (as observed in fire tests and real fires), it is currently recommended that use of this enhancement is limited to slim floor beams and partially encased beams only.

4.6.1 Slim floor beams

A full depth end plate is recommended for beam-to-column connections for *Slimflor* and *ASB* construction because of the torsional loads that may occur during construction. For this type of connection to remain 'simple' or 'flexible' in normal design, the thickness of the endplate is limited to 10 mm. Also, a limitation is imposed on the bolt cross centres.

For a 10 mm thick endplate, the moment of resistance of these connections for various slim floor sections are given in Design Data Sheets 6 and 7. It is assumed that the connections remain at a low temperature, as they are fully encased in concrete, allowing the full strength of the connections to be used for 60 minutes fire resistance. The connection moment is simply added to the allowable moment of resistance of the corresponding steel section (as shown in Design Data Sheets 1 or 2) and compared to the applied moment in fire conditions. The mechanism of utilising the bending resistance of connections in fire is shown diagrammatically in Figure 4.7.

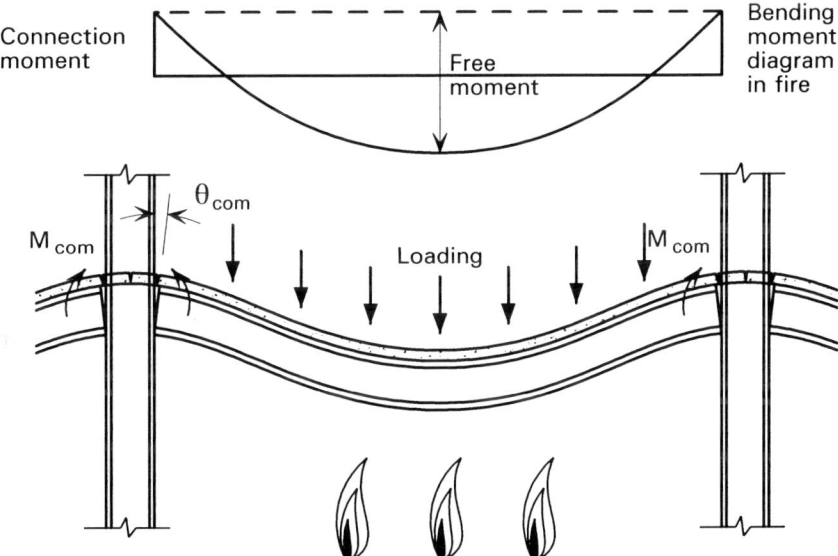

Figure 4.7 *Re-distribution of moment in simply-supported beams in fire conditions*

4.6.2 Partially encased beams

It is not feasible to provide values of moment of resistance for various connections, because of the large number of different connections that can be used. However, the bending resistance of the connection may be calculated from first principles[15]. Provided the connection is encased in concrete, its full resistance can be assumed for 60 minutes fire resistance.

4.6.3 Effect on columns

The resistance of the columns to bending and compression has to be checked for the additional transfer of moment from the beam at the fire limit state. Internal columns are likely to be in a balanced condition, but the design of edge columns may be more critical.

For columns with applied fire protection used in conjunction with unprotected beams, the following simple recommendations are presented. For other types of column the combination of axial load and moments must be checked using the appropriate design rules.

1. If the transfer of moment through the connection at the fire limit state is less than the unfactored moment on the column assumed in normal design (i.e. due to the offset of the connection), then the column should be protected so that its temperature does not exceed 550°C. Depending on the fire protection material used, the required thickness[16] can be specified accordingly.

2. If the transfer of moment through the connection at the fire limit state is greater than the unfactored moment on the column assumed in normal design, then the capacity of the column at the fire limit state should be checked. Alternatively, provided the normal resistance of the column is adequate to resist the additional moment transferred in fire (calculated using unfactored loads), the following recommendations could be used:

    If the transfer of moment through the connection at the fire limit state is not greater than 1.3 times the unfactored moment on the column assumed in normal design, then the column should be protected so that its temperature at the relevant fire resistance period does not exceed 450°C. If the moment is greater, then the amount of protection should be increased. For perimeter columns, brick encasement may often be appropriate.

5 COLUMN DESIGN

This Section presents various methods that can be used to achieve 15, 30 and 60 minutes fire resistance for steel columns. The four types of unprotected and partially protected columns listed in Table 2.1 are covered.

For each type of column, the design methods selected are those considered to be the most appropriate. If more than one method is given then the BS 5950-8 methods are given first. If no reference is made to a particular standard, that type of beam is not included in the standard.

In many cases preliminary design information is given in a Design Data Sheet (Appendix D). These are provided to assist in initial design and designers would be expected to refer to more comprehensive information for final design.

5.1 Effective length of columns in fire

In multi-storey construction sufficient compartmentation is normally provided to ensure that the fire remain confined to one floor level. This means that a column within a fire compartment is continuously connected to a cold column external to the fire affected zone and that column will retain its stiffness. This continuity of the columns will provide a significant degree of rotational restraint to the fire affected column.

In designing columns to BS 5950-8[2], no recognition of improved performance due to end restraint is allowed but the Eurocodes do allow account to be taken of such restraint.

EC3-1-2[4] and EC4-1-2[5] allow some reduction in the effective length of a column provided the column has some rotational end fixity. According to the Eurocodes, where the column at the level under consideration is fully connected to the column above and below, it may be considered to be fully built in at such connections provided that the resistance of the separating elements is at least equal to the fire resistance of the column. Therefore, for columns in an intermediate storey of the building, the buckling length of the column when subjected to fire is 0.5 times the actual column length (system length). For columns on the top storey which are not connected to a column at their upper end, the buckling length should be taken as 0.7 times the system length (Figure 5.1).

In UK National Application Document for EC4-1-2 the factors of 0.5 and 0.7 have been conservatively increased to 0.7 and 0.85 respectively.

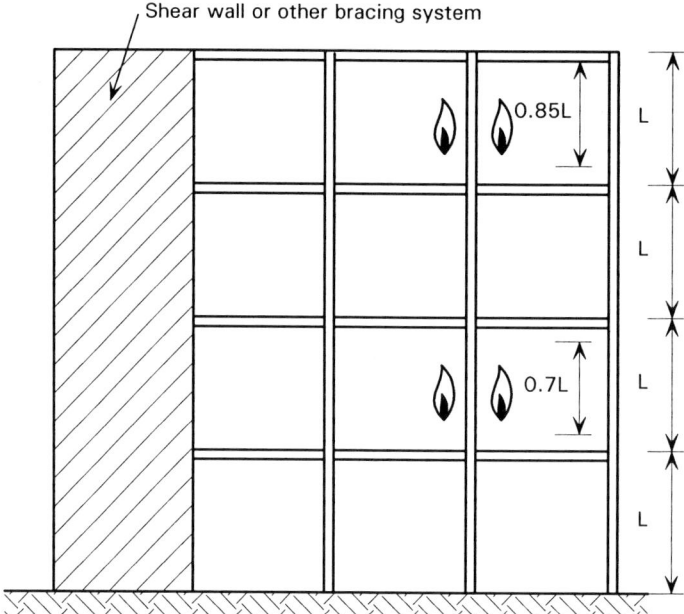

Figure 5.1 *Effective Length of Fire Exposed Columns in Braced Frames. (The values of 0.7 and 0.85 are those in the UK NADs to EC3-1-2 and EC4-1-2)*

5.2 Unprotected columns

Most unprotected Universal Columns would achieve at least 10 to 20 minutes fire resistance in a standard fire test. Very heavy columns may sometimes achieve 30 minutes fire resistance; 60 minutes fire resistance cannot be achieved.

15 minutes fire resistance

In some car park structures, the steel frame is only required to have 15 minutes fire resistance. All Universal Column sections except 152×152×23 UC will achieve at least this fire resistance, as will all sizes of hollow sections. For further details, see Appendix C.

30 minutes fire resistance

Three design methods may be used as follows:

1. Use the 'simple' method in BS 5950-8. Ensure that the load ratio (refer Section 3.33) is less than or equal to 0.6 and that the section factor does not exceed 50 m^{-1}. The latter requirement will result in a minimum section size of 356×406×393 UC.

2. Use the Limiting Temperature method given in BS 5950-8, as explained in Appendix A.

3. Use the design method in EC3-1-2. This involves calculating the Critical Temperature (similar to the Limiting Temperature used in BS 5950-8) based on the applied load at the time of the fire. This Critical Temperature is compared with the maximum steel temperature reached after a given time. This is calculated using a simple differential equation. Realistically, this method requires a simple computer program or spreadsheet.

60 minutes fire resistance

60 minutes fire resistance cannot be practically achieved by unprotected columns, and applied protection should be used.

5.3 Blocked-in columns

The fire resistance of a bare steel column is increased significantly if aerated concrete blocks are placed between the inner faces of its flanges. It is assumed that the blocks do not resist any load but they insulate the web of the steel column and inner surfaces of the flanges, resulting in the steel section increasing in temperature more slowly.

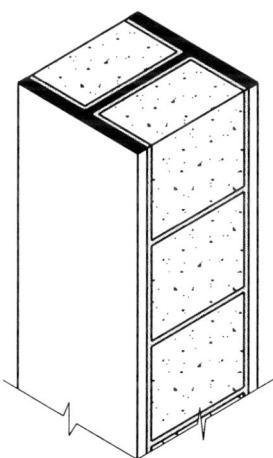

Figure 5.2 *Typical blocked-in column*

30 minutes fire resistance

1. Ensure that the load ratio (refer to Section 3.33) does not exceed 0.6 and that the section factor does not exceed 69 m^{-1}. This limiting section factor value is based on test results conducted by British Steel and the Fire Research Station, which are reported in BRE Digest 317[17]. The minimum size blocked-in column required to obtain a section factor of 69 m^{-1} is a 203×203×52 UC. The concrete blocks should have a minimum density of 475 kg/m^3 in order to ensure adequate thermal capacity and insulating effect.

   (Note: The section factor of 69 m^{-1} is calculated by dividing the exposed surface perimeter of the flanges by the gross section area of the steel column.)

60 minutes fire resistance

Blocked-in columns cannot achieve 60 minutes fire resistance.

5.4 Partially encased columns

The fire resistance of a steel column section can be enhanced by casting concrete between the flanges to form a square section. The concrete shields the web of the column and also acts as a heat sink. Additional reinforcing bars may be included in the concrete encasement in order to increase the column resistance at both the ultimate and fire limit states. Alternatively, the concrete can be unreinforced. In this case, for normal design, the column would be considered as non-composite, but in fire, a limited amount of composite action may sometimes be assumed. The use of unreinforced and reinforced concrete is explained below.

5.4.1 Unreinforced concrete web-infilled columns

30 minutes fire resistance

1. EC4-1-2[5] states that provided the concrete is only used as an insulation material, then no additional reinforcing bars are required and all H section columns will have 30 minutes fire resistance.

60 minutes fire resistance

1. The concrete and steel are required to act compositely to obtain 60 minutes fire resistance. It is assumed that the concrete does not contribute to the normal strength of the column, but is effective at the fire limit state. To ensure load transfer from the steel to the concrete in a fire, web stiffeners must be welded to the top of the column and 'Hilti' shot fired shear connectors (or similar) are fixed to the web of the column at 500 mm centres (Figure 5.3). SCI Technical Report, *The fire resistance of web-infilled steel columns*[18] explains the development of design rules for this type of column. A summary of safe load tables and recommendations from this publication is presented in Design Data Sheet 8

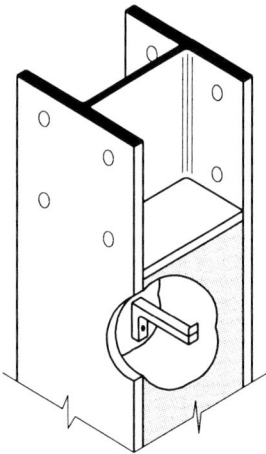

Figure 5.3 *Shear connection in web-infilled column in order to obtain 60 minutes fire resistance*

5.4.2 Reinforced concrete web-infilled columns

30 minutes fire resistance

1. EC4-1-2[5] gives a tabular method for these columns based on a maximum load level, minimum cross-section sizes, minimum axis distance of reinforcing bars and minimum ratio of web to flange thickness (Figure 5.4). The tabular data is incomplete in that the size of the additional reinforcing bars is not given and has to be determined by calculation based on the data given in Annex F of EC4. Alternatively the data developed by SCI can be used. This is presented in Design Data Sheet 9. For 30 minutes fire resistance EC4 states that no additional reinforcing bars are required for a load level of 0.3 or less, based on the composite section. However, nominal reinforcement is recommended to control spalling in fire.

To ensure load transfer from the steel to the concrete, EC4 states that the links should be welded to the web of the column.

60 minutes fire resistance

1. The same method described for 30 minutes can also be applied to 60 minutes fire resistance.

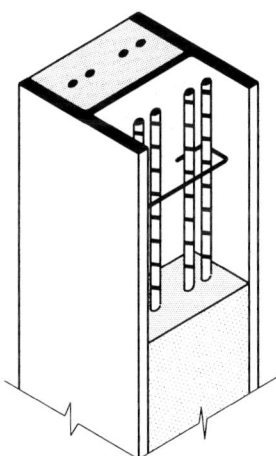

Figure 5.4 *Partially encased reinforced column*

5.5 Concrete filled hollow section columns

Concrete-filled circular, square or rectangular hollow sections are a practical form of composite column achieving a good level of fire resistance. Depending on the required load capacity and fire resistance, the concrete infill may be unreinforced or may be reinforced with bars or chopped steel fibres.

As the section is heated in fire, the outer steel shell loses strength and the load carried by the steel will be redistributed to the concrete core, which loses strength and stiffness more slowly than the steel shell. For greatest efficiency in fire, the composite section should be designed such that the load bearing capacity of the steel is minimised (i.e. using a thinner steel thickness) and the load resistance of the concrete core maximised (i.e. using higher strength concrete and more reinforcement). This will typically result in the overall dimensions of steel sections being larger compared to sections which are intended to have applied fire protection.

Vent holes are required to allow steam from the concrete to escape in fire, in order to ensure that the steel shell will not split. It is recommended that two holes should be provided in opposite faces of the column at 4m centres or at storey height positions, whichever is more frequent. It has been demonstrated that for structural hollow sections up to 400 mm square, two 12 mm diameter vent holes are sufficient.

The presence of a concrete core will increase the effective heat sink of a CHS section and make it act as if it has a lower section factor than that of the hollow section shell.

This means that:

1. All unprotected concrete filled hollow sections columns have an intrinsic 15 minute fire resistance at an ultimate limit state load factor of 0.6 or less, based on the composite section properties.

2. Concrete filled hollow section columns using sections with wall thicknesses greater than 8 mm will have an intrinsic fire resistance of 30 minutes under the same conditions.

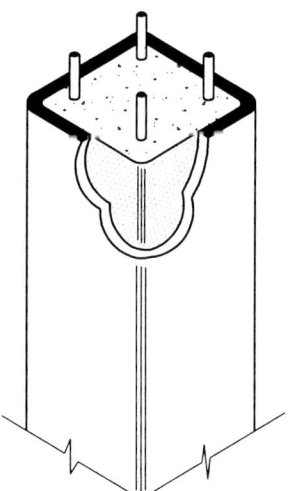

Figure 5.5 *Concrete filled hollow section*

30 minutes fire resistance

The following design methods may be used, as follows:

1. BS 5950-8 can be used for concrete filled square hollow sections not less than 140 mm square, or for rectangular hollow sections not less than 100×200 mm. The concrete core may be unreinforced, bar reinforced, or reinforced with steel fibres. The load ratio is calculated and compared with a tabulated maximum load ratio based on the required fire resistance. This maximum load ratio reduces with increasing fire resistance.

2. A tabular method is presented in EC4-1-2[5], which is based on load level, minimum cross-section dimensions, minimum reinforcing areas and minimum axis distance to reinforcing bars. It is applicable to both square and circular hollow sections.

3. Use the method presented in EC4-1-2 Annex G. This will often result in using less reinforcement than in the tabular method. However, the method presented in Annex G cannot be considered as a simple 'hand calculation' and realistically requires the use of purpose written computer software.

4. SCI design software could be used.

Safe load tables, based on the EC4 method, are presented in Design Data Sheet 10, for typical concrete-filled circular and square hollow sections.

60 minutes fire resistance

The methods given for 30 minutes fire resistance can also be used for the design for 60 minutes fire resistance.

5.6 Brick or block encasement

Masonry walls can offer partial or full encasement to steel columns without the need to use a specialist fire protection contractor. Such walls are commonly used in industrial buildings, and frequently the perimeter columns in low rise multi-storey buildings are protected within brick piers.

Fire resistance tests have shown that a single thickness of brick is sufficient to maintain the temperature of the steel below 150°C. Based on these tests conservative "deemed to satisfy" specifications[19] have been developed which include a requirement for reinforcement in every horizontal joint.

30 minutes fire resistance

A single course of brickwork around a steel section would provide at least 30 minutes fire resistance.

60 minutes fire resistance

A single course of brickwork around a steel section would also provide 60 minutes fire resistance.

6 FLOOR SLAB DESIGN

6.1 Shallow composite floor slabs with steel decking

Composite floors with steel decking have been used for over 20 years in the UK. They achieve fast construction and have good inherent fire resistance due to the continuity provided by the mesh reinforcement. In most circumstances the steel decking can be left unprotected. In fire, the reduced applied loads are resisted by the reinforced concrete action with the exposed steel deck being largely sacrificial. For simple spans or high loads reinforcing bars may be required in the deck ribs in addition to the crack control mesh in the top of the slab.

Composite floors add appreciably to the stability of a building, both for normal design and in fire. The floor acts as a horizontal diaphragm spanning between areas of vertical restraint such as stair wells or braced cores of a building.

30 minutes fire resistance

All composite slabs designed in accordance with BS 5950 or EC4 have at least 30 minutes fire resistance provided the slab thickness is not less than the minimum thickness required to meet the insulation requirements of BS 476[3]. The minimum thicknesses are given in Design Data Sheet 11.

60 minutes fire resistance

The following design methods may be used:

1. Use the simplified design method presented in SCI Publication, *The fire resistance of composite floors with steel decking*[20]. The design consists of a single layer of standard mesh in the concrete, and the method is based on test results and is presented in the form of design tables, as given in Design Data Sheet 11. The method can only be used for slab spans which are continuous over at least one support.

2. Use the design tables issued by most steel deck manufacturers for their products. Tables prepared for manufacturers by SCI are based on the simplified design method mentioned above.

3. Use the methods in SCI publication[20], BS 5950-8[2] and EC4-1-2[5] for calculating the sagging and hogging moment of resistance of composite slabs in fire. The methods take account of the effects of reinforcement. Using tabular data the temperatures of the reinforcing bars and concrete are estimated, and then the moment resistance of the section may be calculated taking into account reduced material strengths at elevated temperatures. Normally, more reinforcement is required than in methods 1 and 2 so it may be a less economic approach. It can be used for slabs which have single or continuous spans.

6.2 Deep composite floor slabs

A more recent development is the deep steel decking which is used in slim floor construction. The first deck of this type was a 210 mm deep steel deck (CF210). This has recently been replaced by a new 225 mm deep deck (SD225). Both decks are supplied by PMF. Deep decks can span up to 6 m (when unpropped), increasing to 8 m when the decking is propped during construction. In fire, the steel deck is again largely sacrificial and therefore individual reinforcing bars of 16, 20, or 25 mm diameter must be placed in each rib to provide the necessary bending resistance. In slim floor construction, the slabs are designed to span as simply-supported between the beams, and no account is taken of the mesh reinforcement over the beams. However, the mesh reinforcement is beneficial in fire conditions.

30 minutes fire resistance

Slabs designed in accordance with BS 5950 or EC4 can achieve at least 30 minutes fire resistance provided the slab thickness is not less than the minimum thickness required to meet the insulation requirements of BS 476[3]. The minimum thicknesses are given in Design Data Sheet 11.

60 minutes fire resistance

The following design methods may be used:

1. Use the PMF published design tables, or the SCI software which was used to develop these tables.

2. Use the methods in BS 5950-8[2] and EC4-1-2[5] for calculating the sagging moment of resistance of composite slabs in fire. Both methods take account of the effect of the reinforcement. Using tabular data the temperatures of the reinforcing bars and concrete are estimated, and then the moment resistance of the section may be calculated taking into account reduced material strengths at elevated temperatures. Fire tests on deep deck systems have shown that these methods predict unconservative temperatures of the reinforcing bars for this type of deck. It is therefore proposed that for SD225 decking, the design values given in Table 6.1 should be used, when calculating the moment resistance of the slab. The corresponding axis distances for the reinforcement are also given. These distances are measured vertically upwards from the bottom of the deck rib.

Table 6.1 *Design reinforcement temperatures for SD225 steel decking*

Fire resistance (mins.)	Recommended axis distance (mm)	Reinforcement temperature (°C)
60	70	398
90	70	598
120	100	586

6.3 Precast concrete slabs

Most precast concrete units used in steel construction are hollow core units. These can have nominally up to 240 minutes fire resistance. They are used in shelf angle construction and slim floor construction.

For 30 minutes fire resistance, no special detailing is required, but for 60 minutes a continuous reinforced structural topping should be used with the ends of the hollowcore units filled for a length equal to the depth of the hollow cores. If a structural topping is not provided, then reinforcement should be provided that is continuous from one unit to the next. For slim floor construction this reinforcement may pass either through the web of the beam or over the top of the beam, and should be embedded within the joints between the units.

Further guidance may be obtained from the Precast Flooring Federation or from the supplier of the hollow core units.

7 STEEL MEMBERS SUPPORTING COMPARTMENT WALLS

7.1 Thermal bridging

Steel members built into compartment walls must not cause the wall to lose its ability to act as a separating element and thus allow fire to spread. Unprotected steel may act as a hot bridge which would affect the ability of the wall to meet the "insulation" criteria set out in BS476[3] which states that the temperatures on the unexposed side of the wall must not rise by more than an average of 140°C, with no part of the wall exceeding 180°C, during the fire period.

This publication is primarily concerned with the fire resistance of different types of unprotected steel. However, when an otherwise unprotected beam or column forms part of a wall or floor it may sometimes require some applied fire protection so that the wall or floor still fulfills its separating function. Recommendations for fire protecting steel beams and columns in the plane of a wall or crossing a wall are given in Table 7.1. The protection recommendations are based on the insulation criterion of BS 476 which apply to separating elements. Generally, to meet these requirements, only nominal protection is needed. Any protection required for the section to maintain its load carrying ability *must* be separately assessed.

7.2 Effect of structural deformation

It is difficult to give rules for the details that should be adopted at the top of compartment walls to allow for the deformation of the structure in a fire. This is because of the difficulty in quantifying the deformations which may occur in a fire, when individual members act as part of a frame. The deformations of these members are almost certainly less than those that are measured in a standard fire resistance test, when members are tested simply supported and in isolation. The experience from real fires, in which the members had the appropriate fire resistance (in terms of furnace performance), is that deformations are quite small. This is in part due to the framing effect of other members of the structure, including the floors.

Additionally, the applied loading at the time of a fire will rarely be the design maximum load which the member is expected to resist after 30 or 60 minutes fire exposure. Studies have shown that the general levels of loading in buildings are quite low and, even though reduced factors are used to calculate the load at the fire limit state, the actual loads are almost certainly lower than those assumed in design.

The details needed depend on the type of wall and its boundary conditions. Masonry walls, even if they are not designed to be load bearing, have an appreciable compressive resistance and can often support a beam that deforms in fire. In view of this, no special recommendations are necessary for masonry walls, but guidance on lightweight walls is appropriate.

Table 7.1 *Recommendations to eliminate thermal bridging through steel members within compartment walls*

Beam or column detail		Fire resistance (mins)	
		30	60
Conventional beam in plane of wall		Beam must be protected**	Beam must be protected**
Slim floor beam in plane of wall		No local protection required provided wall at least 100 mm thick.	No local protection required provided wall at least 140 mm thick.
Asymmetric beam or partially encased beam in plane of wall		No local protection required provided wall at least 100 mm thick.	Protection must be applied to underside of flange.
Conventional beam crossing wall		Local protection required	Local protection required
Slim floor or Asymmetric beam or partially encased beam crossing wall		No protection required provided wall at least 140 mm thick.	No protection required provided wall at least 200 mm thick.
Conventional column in wall		*Lightweight wall*: No protection required provided wall at least 100 mm thick	*Lightweight wall*: Column must be protected
		Masonry wall No protection required provided wall at least 100 mm thick	*Masonry wall* No protection required provided wall at least 150 mm thick
Conventional column in wall		Column must be protected.	Column must be protected
RHS column in wall or Partially encased column in wall		*Lightweight wall*: No protection required provided wall at least 140 mm thick	Lightweight wall: No protection required provided wall at least 200 mm thick
		Masonry wall: No protection required provided wall at least 100 mm thick	*Masonry wall*: No protection required provided wall at least 140 mm thick
Partially encased column in wall		No protection required	No protection required provided wall at least 100 mm thick and section depth at least 200mm

In all cases the wall is assumed to have 30 or 60 minutes fire resistance. Where a masonry wall is specified it is assumed not to be in thermal contact with the beam.

Generally protection, is only required to be nominal (10 mm of fire protection material). However, for cases marked ** protection should be based on exposed A/V.

Where a beam crosses a wall, any protection should extend 150 mm either side of the wall.

Intumescent coatings are not acceptable to meet compartment wall insulation requirements.

Lightweight stud walls are very weak and offer little resistance and should always be constructed with some form of deflection head. This will allow the structure above the wall to deflect downwards without compressing the wall. In cases where the beam is in line with the wall, the beam deflection will be small as it is only heated from one side. In this case, the allowance for deflection used in normal design should be adequate. When the beam spans across the wall, a more severe condition exists, as described below.

For beams which have an inherent fire resistance of 60 minutes, such as slim floor beams, no special provision needs be made for use in buildings required to have 30 minutes fire resistance, as their deflections will be small. In other cases, for beams of up to 6 m span, a 25 mm allowance is recommended, increasing to 50 mm for a 9 m span. For the range of beams described in this publication, the span of a beam without applied fire protection is practically limited to about 9 m but for longer spans it is probably more appropriate to limit the deflection allowance to 50 mm (as a practical limit) and instead to over design the beam in some way at the fire limit state. This will have the effect of reducing the load ratio and thus will reduce deflections.

8 OVERALL FRAME STABILITY

The overall stability of steel or composite frames has to be maintained at the fire limit state. This will depend on the mechanism adopted to resist lateral loads at the ultimate limit state.

8.1 Braced frames

If concrete walls have been constructed to enclose lift, stair and service cores and these are also used to provide lateral stability, then no additional checks for stability in fire conditions are required, provided that the minimum cross-section sizes and cover to reinforcement for a given fire resistance period are used, as specified in BS 8110[21].

If braced bays are used to provide lateral stability then the steel members in these areas will require fire protection, unless it can be shown that alternative means of resisting lateral loads exist in fire conditions. One way might be to create a sway frame utilising the connection stiffness.

Forces are transferred to the core areas by the diaphragm action of the floor slab. If a composite floor is used, then, diaphragm action is ensured up to the fire resistance period of the floor. If precast concrete floors are used, a structural topping is recommended to provide floor diaphragm action.

8.2 Sway frames

Although steel loses strength and stiffness in a fire, it is initially the loss of stiffness that is most noticeable. In a frame, if all the columns at a particular level are affected by the fire, the loss of sway stiffness could be the cause of failure. Collapse in sway or unbraced frames could occur due to the build up of "P-delta" effects, caused by the deflection of the columns under vertical load. However, it can be argued that in a fire it is unlikely that all the columns and associated beams on one level will be affected by the fire at the same time, and it is likely that sufficient resistance to horizontal forces will be provided during a fire. Although this argument is almost certainly true, it is difficult to guarantee or quantify this effect at the fire limit state. For design purposes therefore, the horizontal strength and stiffness of sway frames must be considered in fire. This may be achieved by ensuring that certain sub-frames are fire protected or contained within a protected shaft that provides the necessary resistance.

Alternatively, BS 5950-8[2] specifies a conservative Limiting Temperature of 520°C for sway frames, which takes account of additional "P-delta" effects.

9 EXTERNAL STEELWORK

EC3-1-2, Annex C[4] and EC1-2-2[6], contain guidelines for the calculation of the temperature rise of steelwork located outside the building envelope. The calculation is based on the premise that external steel columns and beams may be partially shielded from direct exposure to the fire plume discharging through openings, with the rise in steel temperatures being caused by the radiant heat from the flame. The design strategy relies on the use of robust cladding which does not fail or radiate heat in fire. This design method can justify the use of unprotected external steel, and has been applied successfully on a number of steel framed buildings.

The principles in the Eurocodes are based on work carried out in the UK and France. These principles are explained in SCI publication *Fire safety of bare external structural steel*[22].

10 CONNECTION DETAILS

Most readers will be familiar with the more common types of connection used in steel framed buildings. However, some of the combinations of beam and column that might result from the use of unprotected steel require what could be considered as non-standard connections. Suggestions on some specific types of connection are given in the following Sections.

For *Slimflor* beams, the connections illustrated may not meet the dimensional requirements shown in Design Data Sheets 6 or 7. In that case, if the moment resistance of the connection is important to the design then it must be assessed from first principles.

10.1 General principles

In connecting unprotected sections the following principles should be followed.

1. For columns, the compressive load path must be maintained.

    For web-infilled columns (see 5.4.1), the connection area does not have to be concreted but it does have to be shielded. This may be achieved either with the floor slab or by some form of locally applied protection. In all other cases, the reinforced concrete must be continuous through the connection.

2. Bolts should be either:

    a) Embedded in the floor slab.
    or, b) Shielded from the effects of fire.
    or, c) Considered to be sacrificial in fire. The rest of the bolt group should be designed to carry the reduced loads appropriate to the fire limit state.

10.2 Connection of internal *Slimflor* beams to columns

The connection between a *slimflor* beam and a column is best achieved by using full depth end plates, welded to the beam and bolted to the column. The connection detail will depend on the type of column.

A full-depth end plate is required for *slimflor* beams to resist torsional loads due to out-of-balance loading during construction (refer to SCI publications[8][9]).

The connections illustrated show Fabricated *Slimflor* Beams. Connections for Assymetric *Slimflor* Beams are generally similar.

10.2.1 Connection to column with applied protection

For connections to the major axis of the column, a simple end plate is welded to the end of the beam and then bolted to the column flange. For connections to the minor axis of the column, a similar detail may be used. The beam flanges may be notched to fit between the column flanges. The fire protection to the columns, which is generally board protection, can then be applied in the normal way.

10.2.2 Connection to partially encased columns

For unreinforced concrete web-infilled columns, it is not necessary for the concrete to be continuous through the connection zone, but it is important that the unconcreted portions are effectively shielded from any fire. In the case of a slim floor construction much of this area is encased by the floor slab.

The concrete between the flanges of the column must be contained by a stiffener at the top of the column (below the floor). The concrete between the flanges above the floor does not need to be contained by a stiffener provided it can bear, in some way on the floor slab or the incoming beam. However, providing a stiffener at this location will ease the placement of concrete (which will typically be carried out before erection), thus eliminating the need for additional formwork.

For connection to the minor axis two options are illustrated in Figure 10.1. If the concrete infill to the column is placed after erection the beam may be bolted directly to the web of the column. If the concreting is carried out prior to erection a plate may be welded across the column flanges and a conventional end plate connection can then be made. To allow access for bolting in the latter case, the concrete infill must be omitted for approximately 200 mm above the connection. This area must be subsequently filled with concrete. It is important to ensure load transfer from the concrete infill to the beam or floor.

Connections to the major axis may be made in the conventional way by bolting through the flange.

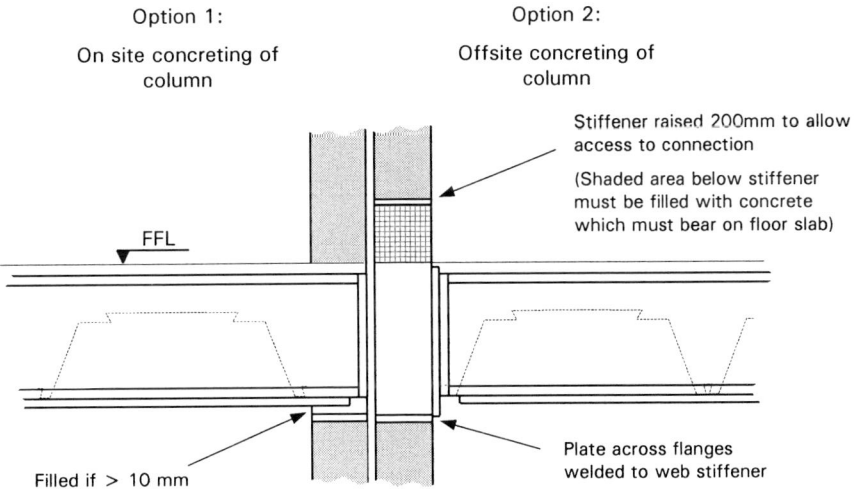

Figure 10.1 *Two options for connecting slimflor beams to the minor axis of web-infilled columns*

For reinforced partially encased composite columns, the concrete infill and bar reinforcement should be continuous through the connection. A fin plate connection is unsuitable for this situation because a slim floor beam requires some form of torsional restraint, generally resulting in the need for a full-depth end plate.

It is normal to place the concrete with the column horizontal and also for the concrete to be cast continuous over more than one storey height. One solution to the connection to the minor axis is to weld a plate across the toes of the column that extends out either side. The slim floor beam would be fitted with a similar width end plate and the connection can therefore be made with no interaction from the concrete. A similar detail will often be suitable for connection to the strong axis of the column.

There may be occasions when a wide plate may cause problems with connections to the other axes and another solution is to weld a short section of RSJ to the web of the column and make a normal end plate connection outside the column. The narrow RSJ becomes embedded in the concrete and can therefore be made from an appreciably smaller and lighter section than the *slimflor* beam.

Both options are illustrated in Figure 10.2

For connection to the major axis wide plates, welded cleats or fin plates must be used.

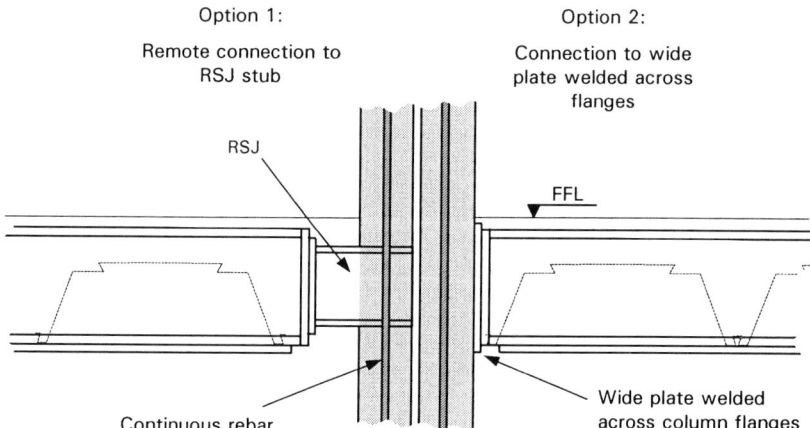

Figure 10.2 *Two options for connecting of a slimflor beam to the minor axis of a reinforced partially encased column*

10.2.3 Connection to concrete filled hollow section columns

Connection to concrete filled hollow sections may be achieved using the Flowdrill technique, Hollo-bolts (a form of expanding bolt), or by plates welded to the column.

Flowdrill is a method of making a threaded hole in the wall of a hollow section. If the concrete infill is placed on site, then all bolts need to be placed before this operation is carried out. However, if the sections are filled under factory conditions, greased bolts must be inserted in the threaded holes which are then removed after concrete filling. This allows the bolts to be fitted during erection.

Hollo-bolts can only be used if the concrete in-fill is placed on site and must be installed before filling.

If the connection is formed using end plates or fin plates then the plates will be embedded in the floor slab and thus shielded from fire. Generally, the bolts are embedded in concrete and thus shielded from the effects of the fire.

Alternatively, bolts could be made to pass completely through the section but these bolts could interfere with internal reinforcement and are also an expensive form of construction.

10.3 Connection of RHS *Slimflor* edge beam to columns

In slim floor construction, a rectangular hollow section (RHS) with a welded plate may be used as an edge beam (see Section 4.3). The design of RHS *Slimflor* edge beams, at the fire limit state, is based on the assumption that the outer face of the beam is shielded from fire by the cladding system and the inner face by the slab.

Fire stopping between the outside face of the RHS and the cladding should normally be provided at the level of the bottom of the RHS. If it is provided at floor level, then the outside face of the RHS must be fire protected.

10.3.1 Connection to conventionally protected column

For connections to the major axis of the column, a simple end plate is welded to the end of the RHS beam and then bolted to the column flange in the same way as with an ordinary Fabricated *Slimflor* Beam.

For connection to the minor axis of the column, a similar detail may be possible if the column is sufficiently large. Normally, it will be necessary to weld a plate across the flange tips. Using this method, the RHS may be located offset from the column and flush with the column flange (Figure 10.3). The fire protection (usually board) can then be applied to the column in the normal way.

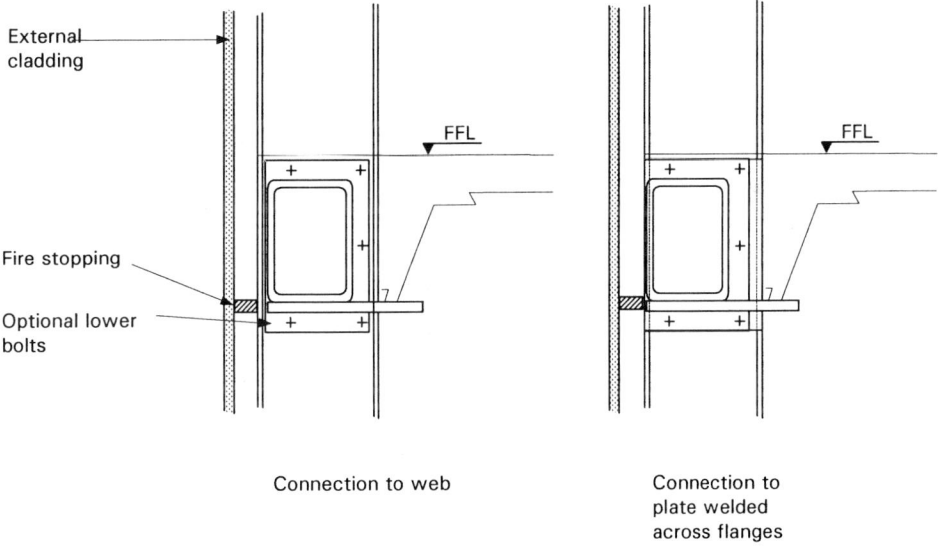

Figure 10.3 *Two options for connecting RHS Slimflor edge beams to the minor axis of a conventionally protected column*

10.3.2 Connection to partially encased columns

For unreinforced partially encased columns, where the concrete is not required to be continuous, connections to the minor and major axes of the column can be detailed as shown in Figure 10.4.

For reinforced partially encased columns, the same connection details as for an internal *slimflor* beam can be adopted.

10.3.3 Connection to a concrete filled hollow sections

This connection requires either the use of the Flowdrill technique or Hollo-bolts or welded plates (Figure 10.5). If welded plates are used the parts of the connection on the outside of the building must be shielded from fire.

Bolts positioned below the floor slab must either be fire protected in some way or may be treated as sacrificial. In fire, the strength of unprotected bolts must be discounted and the remaining bolts designed to carry the reduced loads.

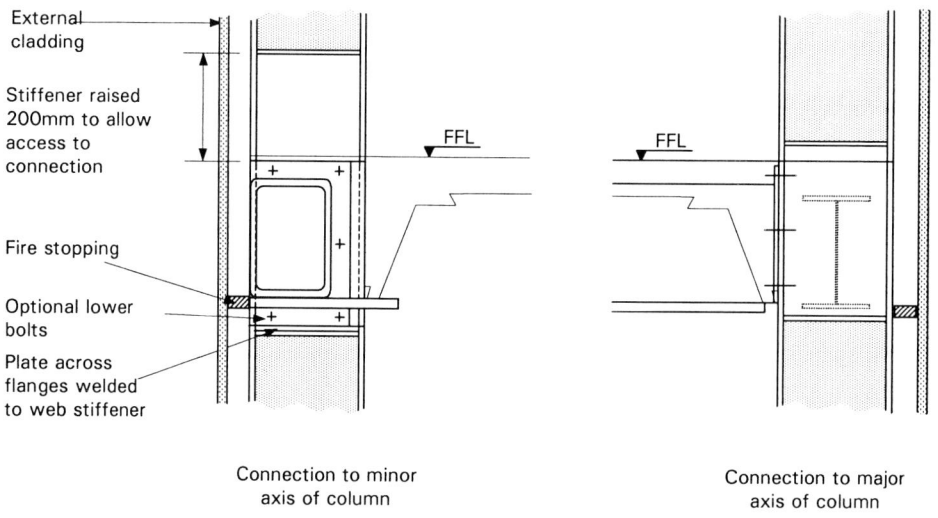

Figure 10.4 *Connections of RHS slimflor edge beam to web-infilled column*

10.4 Shelf angle floor beam

Shelf angle floor beams can be connected to other beams or columns using either welded end plates or fin plates (Figure 10.6). In both cases it is likely that some bolts will be exposed and some bolts will be encased in concrete or grout within the floor slab. In fire, the resistance of the exposed bolts must be discounted, and the encased bolts must be capable of supporting the reduced shear load.

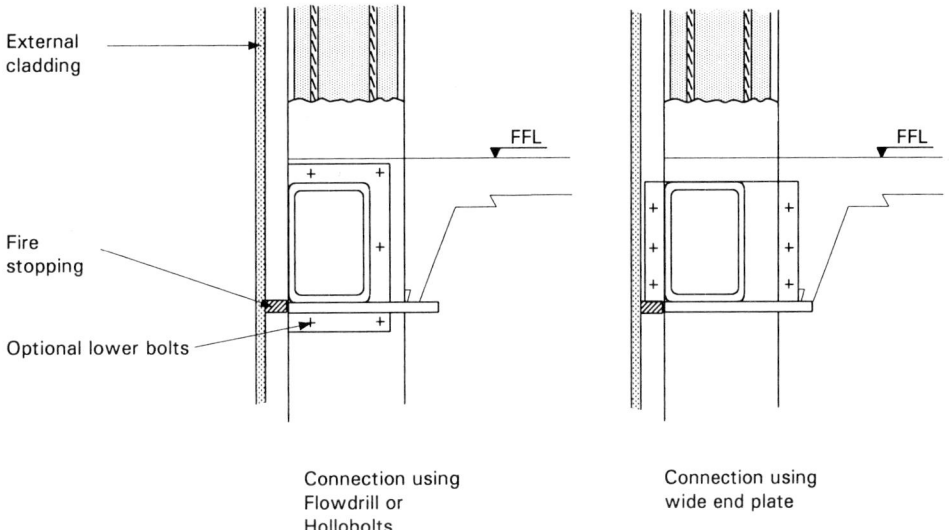

Figure 10.5 *Connections of RHS slim floor edge beam to concrete filled RHS column*

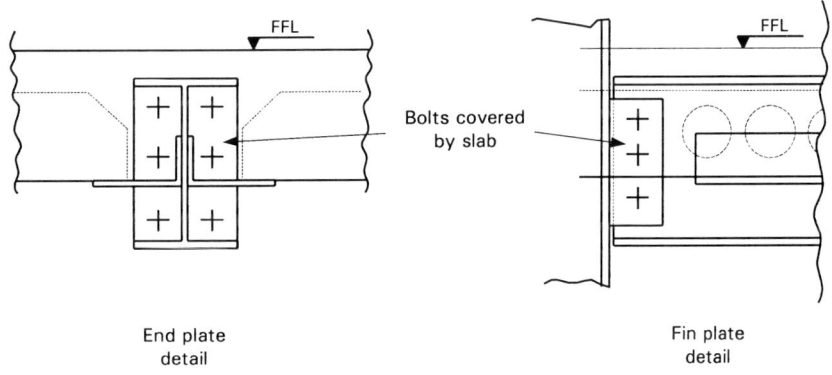

Figure 10.6 *Connection of shelf angle beam using end plate or fin plate*

10.5 Partially encased beams

For these types of beams the connection areas are fire protected by encasing after erection of the structure. The connection detail should be such that later encasement of the connection is feasible. A common method is to cut away the top flange as shown in Figure 10.7. The concrete in the connection zone is then placed when the slab is cast.

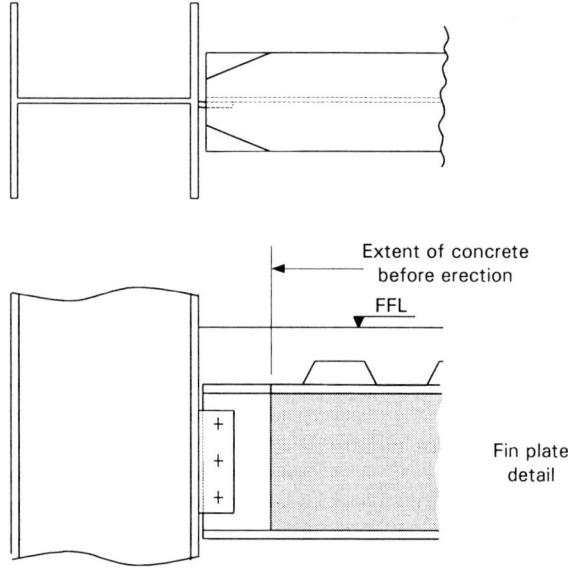

Figure 10.7 *Fin plate connection of partially encased beam*

Fin plates are most commonly used for beam-to-beam connections as shown in Figure 10.8. Again, the top flange is cut to aid concrete encasement of the connections following erection.

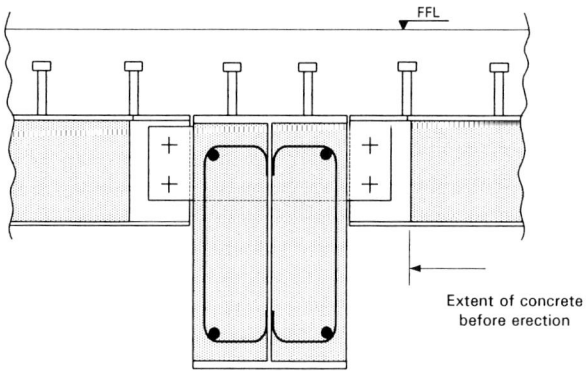

Figure 10.8 *Beam-to-beam connection for partially encased beams*

11 DESIGN EXAMPLES

The use of the structural fire engineering techniques is best illustrated by design examples. The examples given here are based upon a 4 storey office building (Figure 11.1) with a plan of 48 m by 12 m. The storey height is assumed to be 4 m. The idealized building plan is shown in Figure 11.2. The required fire resistance is assumed to be 60 minutes. The design loading is taken as 3.5 kN/m^2 plus 1.0 kN/m^2 for partitions, and 0.7 kN/m^2 for ceiling services and raised floor. Means of escape are provided via stairs which are enclosed in protected shafts at either end of the building.

As this guide is concerned with design for fire conditions the basis for the normal design of the members will not generally be discussed. In some cases, the section sizes selected may not meet all the normal design requirements and should be taken as illustrative only.

Figure 11.1 *General view of the building used in the design examples*

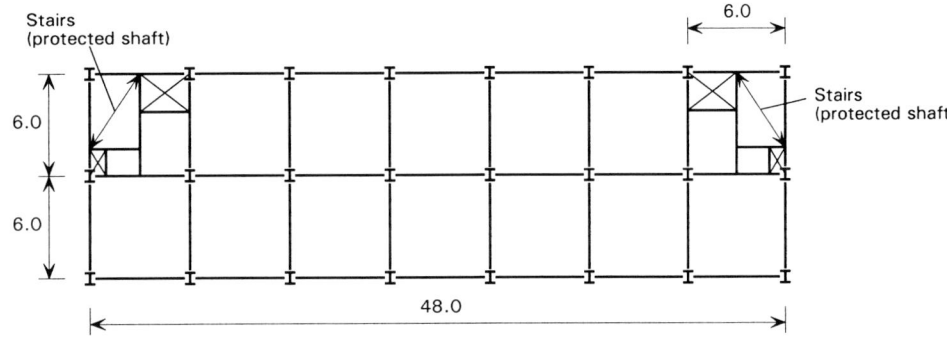

Figure 11.2 *Structural layout of the building used in the design examples*

11.1 Beam design

The following beam types are considered:

1. *Slimflor* floor beam (UC with welded plate)
2. ASB (Asymmetrical Slim Floor Beam)
3. Shelf angle beam
4. Partially encased beam.

11.1.1 Slimflor beams (UC with welded plate)

The general form of a *Slimflor* beam is shown in Figure 11.3.

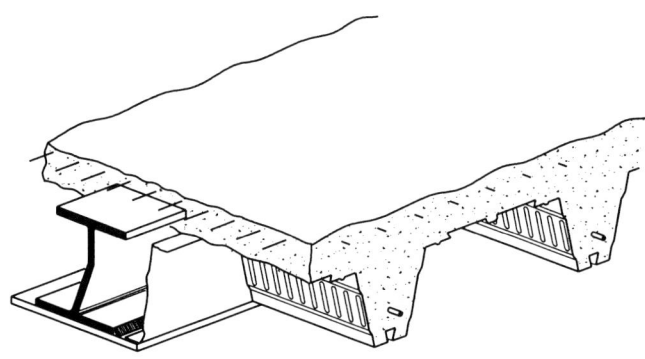

Figure 11.3 *Typical Slimflor beam*

The floor slab may be constructed from precast concrete units or using deep composite decking. In this example, the slab has a span of 6 metres and is assumed to be constructed using a composite SD225 deep deck. For the design of *Slimflor* beams using precast floors, reference to SCI publication *Slim floor design and construction*[8].

For 60 minutes fire resistance, the depth of the floor slab must comply with the insulation requirements of BS 476 Part 20[3]. If lightweight concrete is used, the minimum depth of concrete above the steel decking is 60 mm (see Design Data Sheet 11). The slab depth is therefore 285 mm (225+60mm).

The total weight of composite slab is 2.55 kN/m^2

Slimflor beams are normally designed using SCI software or design tables. (See Design Data Sheet 1). If the design tables are used, then the flange plate thickness must not be less than 15 mm thick.

For *Slimflor* beams, the fire design is based on the maximum applied moment on the beam. The load at the fire limit state is calculated using the reduced partial load factor of 0.8 for non-permanent loads, as given in BS 5950-8[2]. The partitions and raised floor and services are normally classed as permanent imposed loads with a partial load factor of 1.0.

Table 11.1 *Design load summary for Slimflor beams*

Loading	Characteristic load (kN/m^2)	Partial load factors (BS 5950-8)	Design load in fire (kN/m^2)
Self weight of composite slab	2.55	1.0	2.55
Self weight of beam	0.4	1.0	0.4
Raised floor, ceiling and services	0.7	1.0	0.7
Partitions	1.0	1.0	1.0
Imposed load	3.5	0.8	2.8
Total factored load at fire limit state			7.45

The design moment at the fire limit state is therefore:

$7.45 \times 6.0 \times 6.0^2 / 8 = 201.2$ kNm

From Design Data Sheet 1, for non-composite construction:

Use $203 \times 203 \times 52$ UC with 15 mm plate in S355 steel
(moment resistance = 207 kNm).

Slimflor beams may also be designed as composite beams with shear connectors welded to the top flange. In this case the depth of concrete above the top flange of the steel section must be at least 85 mm. The slab depth should be increased to 291 mm in order to achieve the 85 mm cover required for placing the shear connectors.

From Design Data Sheet 1, for composite construction:

Use $203 \times 203 \times 52$ UC with 15 mm plate in S355 steel
(Moment of resistance = 242 kNm).

It can be seen that for the loading in this example, no advantage has been gained by designing the beams as composite. This is because at the fire limit state some composite action due to bond is assumed in the design of slim floor beams that are otherwise treated as non-composite in normal design. However, from Design Data Sheet 1, it can be seen that, in general, at the fire limit state, as the section increases in size, a noticeable saving can be obtained if the beam is designed as composite in fire conditions.

11.1.2 Asymmetric *Slimflor* beam (ASB)

This type of section has been developed to be used specifically with SD225 deep decking. The beam is designed to act compositely with the encased concrete at the ultimate, serviceability and fire limit states. Composite action is achieved by the bond between the concrete and steel section, which is enhanced by the raised rib pattern on the top flange. A minimum 30 mm thickness of concrete must be provided over the top flange if composite action is to be developed. The general form of the Asymmetric *Slimfloor* Beam (ASB) is shown in Figure 11.4.

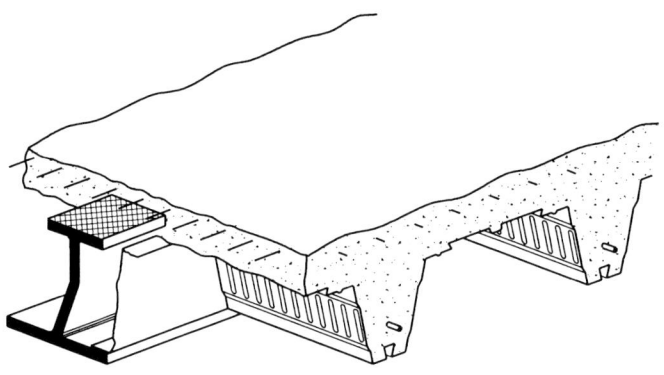

Figure 11.4 *Asymmetric Slimflor Beam and deep composite slab*

Table 11.2 *Design load summary for Asymmetric Beam*

Loading	Characteristic load (kN/m^2)	Partial load factors (BS 5950-8)	Design load in fire (kN/m^2)
Self weight of composite slab	2.75	1.0	2.75
Self weight of beam	0.25	1.0	0.25
Raised floor, ceiling and services	0.7	1.0	0.7
Partitions	1.0	1.0	1.0
Imposed load	3.5	0.8	2.8
Total factored load at the fire limit state			**7.5**

The design moment at the fire limit state is therefore:

$7.50 \times 6.0 \times 6.0^2 / 8 = 202.5$ kNm

From Design Data Sheet 2:

Use 280 ASB 100 section in S355 steel
(Moment resistance = 266 kNm).

11.1.3 Shelf angle beam

The general form of a shelf angle beam is shown in Figure 11.5. The precast hollow core units are assumed to have a 40 mm structural screed, and a span of 6 metres. The beams are assumed to be 406×152×60UB in S355 steel, with two 125×75×12 angles in S355 steel. The angles are not considered to contribute to the bending resistance of the beam in normal design.

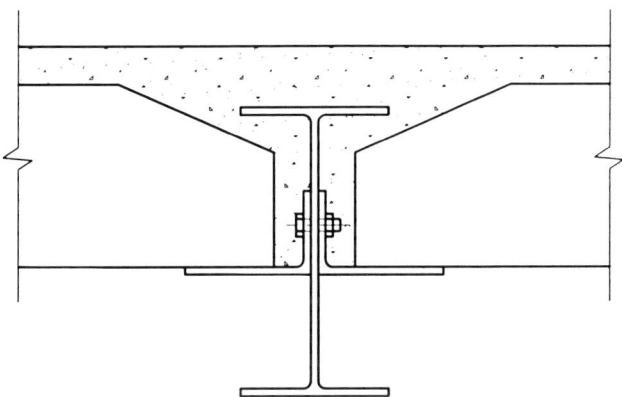

Figure 11.5 *Typical shelf angle beam supporting precast concrete slabs*

Table 11.3 *Design load summary for shelf angle beam*

Loading	Characteristic load (kN/m^2)	Partial load factors (BS 5950-8)	Design load in fire (kN/m^2)
Self weight of precast floor slab	4.7	1.0	4.7
Raised floor, ceiling and services	0.7	1.0	0.7
Partitions	1.0	1.0	1.0
Imposed load	3.5	0.8	2.8
Total factored load at the fire limit state			**9.2**

The design moment at the fire limit state is therefore:

$9.2 \times 6.0 \times 6.0^2 / 8 = 248.4$ kNm

The design of shelf angle floor beams in fire is governed by Appendix E of BS 5950-8. The two angles are included in the bending resistance calculation for fire. The plastic moment resistance is calculated assuming a temperature distribution for the fire resistance period under consideration. The net moment resistance increases as the angles are placed lower on the section, and more of the section is insulated by the concrete floor. However, this has the disadvantage of increasing the slab depth.

For simplified design, the table in Design Data Sheet 4 may be used. The use of this table is explained below.

For the normal, cold, design, the 457×152×60 UB (in S355 steel) has a moment resistance of 456 kNm. The load ratio to be achieved is therefore 0.55 (i.e 248/456). From the table, by linear interpolation, the angle position, H, to achieve this is 210 mm. If the structural screed above the beam is 40 mm the depth of the hollow core unit plus the depth of the screed must be at least 250 mm. The weight corresponding to this depth of slab should not exceed that initially assumed in the design loads.

11.1.4 Partially encased beam

The general form of a partially encased beam is shown in Figure 11.6.

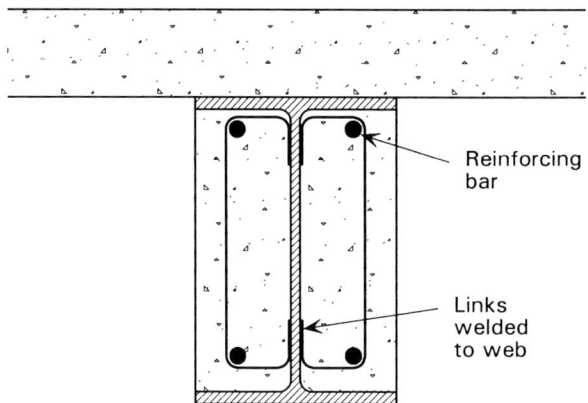

Figure 11.6 *Typical partially encased beam supporting a concrete slab*

For this design option, the floor is assumed to be a composite floor of 130 mm depth. The slab span is 3 metres between secondary beams. The floor beam layout is shown in Figure 11.7.

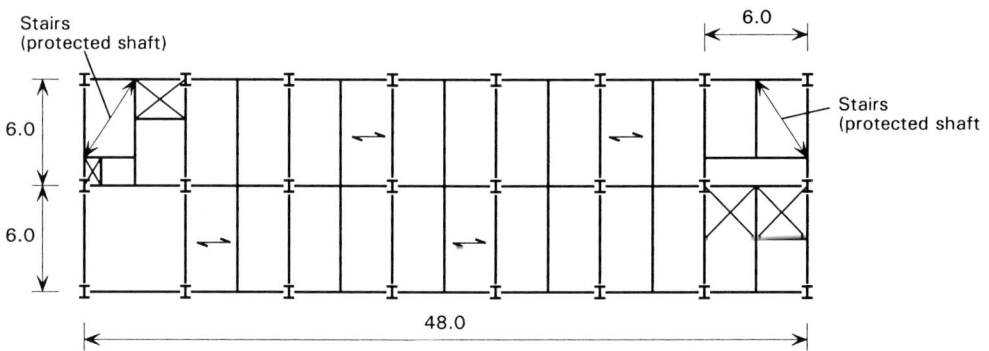

Figure 11.7 *Structural plan layout for floor span of 3 m*

The beam is assumed to act compositely with the slab and is designed for fire in accordance EC4-1-2[5]. EC4 presents two alternatives; a simple tabular method, and a more complex method, which realistically is only suitable for computer use.

The design loads are similar to those in the earlier example.

Table 11.4 *Design load summary for partially encased beam*

Loading	Characteristic load (kN/m²)	Partial load factors (BS 5950-8)	Design load in fire (kN/m²)
Self weight of composite slab	2.4	1.0	2.4
Self weight of steel beam and concrete	0.3	0.7	0.3
Raised floor, ceiling and services	0.7	1.0	0.7
Partitions	1.0	1.0	1.0
Imposed load	3.5	0.8	2.8
Total factored load at the fire limit state			**7.2**

The design moment of the beam of 6.0 m span at the fire limit state is:

$7.2 \times 3.0 \times 6.0^2 / 8 = 97.2$ kNm

For preliminary design the design table in Design Data Sheet 5 may be used. The Part of the table concerning 30 and 60 minutes fire resistance is also reproduced in Table 11.1. The table gives a minimum width and height of section together with minimum area of reinforcement for a given load level and fire resistance. From the table it can be seen that for load levels up to 0.5 and 60 minutes fire resistance, the concrete requires no longitudinal reinforcement (although a minimum area of reinforcement should be specified to control spalling).

For this design example, use a $254 \times 102 \times 25$ UB in S355 steel acting compositely with the floor slab. The moment resistance at the ultimate limit state is approximately 195 kNm, assuming 40% shear connection. This results in a load ratio of $97.2/195.0 = 0.5$ at the fire limit state.

The beam must meet the dimensional requirements of EC4-1-2. The information in Design Data Sheet 5 is taken from EC4. An extract is given in Table 11.5. The width of the beam flange is 101.9 mm and the depth of the beam is 257.2 mm. The use of the EC4-1-2 design table is not straightforward so the following procedure is suggested:

1. Calculate the load level. However, the tabular data is for values of load level of 0.3, 0.5 and 0.7 only and EC4-1-2 states that linear interpolation is permissible. This involves interpolating values of "min b" and reinforcement ("min b" is the minimum allowable flange width).

2. Compare the actual flange width with the "min b" value in the table starting with the smallest value of "min b". If the width is less than "min b" the beam cannot achieve the required fire resistance and another section must be selected.

3. For this value of "min b", check if the beam depth meets the corresponding requirement for "h". The beam will have the required fire resistance if the condition is met.

4. Repeat steps 2 and 3 using the next, larger value of "min b" until both the beam width and depth meet the requirements.

In the example the beam depth cannot satisfy the requirement "h ≥ 3.0×min b", because the depth is inadequate (less than 300 mm), nor can it satisfy the requirement "h ≥ 2.0×min b", because the width is less than "min b" (120 mm) so, according to EC4-1-2, 60 minutes fire resistance cannot be achieved.

Increasing the beam size to 254×146×31 does allow 60 minutes fire resistance to be achieved because "h ≥ 2.0×min b" requirement can be met.

Table 11.5 *EC4-1-2 design table for partially encased composite beams*

Load level (η)	Minimum cross-sectional dimensions (mm) and reinforcement for periods of fire resistance of :			
	R30		R60	
	min b	reinforcement	min b	reinforcement
$\eta = 0.3$				
h ≥ 0.9 × min b	70	0.0	100	0.0
h ≥ 1.5 × min b	60	0.0	100	0.0
h ≥ 2.0 × min b	60	0.0	100	0.0
$\eta = 0.5$				
h ≥ 0.9 × min b	80	0.0	170	0.0
h ≥ 1.5 × min b	80	0.0	150	0.0
h ≥ 2.0 × min b	70	0.0	120	0.0
h ≥ 3.0 × min b	60	0.0	100	0.0
$\eta = 0.7$				
h ≥ 0.9 × min b	80	0.0	270	0.4
h ≥ 1.5 × min b	80	0.0	240	0.3
h ≥ 2.0 × min b	70	0.0	190	0.3
h ≥ 3.0 × min b	70	0.0	170	0.2

Notes:

η denotes load level

b denotes width of section

h denotes depth of section

The amount of reinforcement is expressed as a proportion of the area of one flange

Reinforcement is required for heavily loaded beams or for more than 60 minutes fire resistance. For the design example only a minimum amount of reinforcement is required to control spalling.

11.2 Column design

Design examples for the following types of column are presented:

1. Partially encased column (unreinforced web-infilled column).
2. Concrete filled structural hollow section

From the floor layout (Figure 11.2) it can be seen that there are three main column types: central, edge, and corner column. For this design example, an edge column will be considered.

11.2.1 Unreinforced web-infilled column

The general form of a web-infilled column and required shear connection is shown in Figure 11.8.

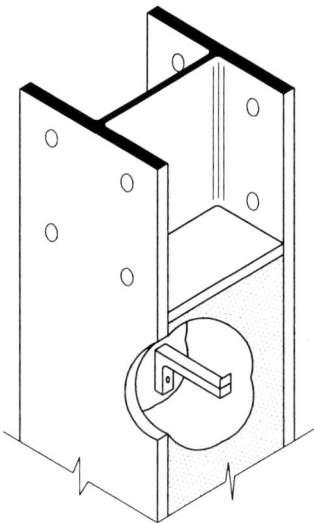

Figure 11.8 *Web-infilled column showing required shear connection*

Web-infilled columns are designed as bare steel columns (using the simple design rules to BS 5950-1); the effects of composite action are only considered in fire.

The design loads for normal (cold) design for the bottom storey column are:

Characteristic loads on the column: Dead   351 kN
                                     Imposed 202 kN

In addition, moments due to out-of-balance loading causing bending about the major axis of the column are:

Dead load moment   8.6 kNm
Imposed load moment   5.2 kNm

At the fire limit state the loads are:

Compression load = 351 + 0.8 × 202 = 513 kN
Moment   = 8.6 + 0.8 × 5.2 = 12.8 kNm

According to the UK NAD for EC3-1-2[4], the effective length of the column may be assumed to be reduced for fire design and taken to be 0.7 of the system length, which leads to an effective column length of 2.8 m. For this type of column, the fire design may be based on the SCI publication *The fire resistance of web-infilled steel columns*[18], and the summary tables presented in Design Data Sheet 8. In the tables, the compressive load and moment resistances are given, and these may be combined using a simple linear interaction formula. These tables give the maximum compressive load in terms of the effective length of the column used in normal design, with the 0.7 factor taken into account when compiling the tables.

From the tables, for S355, steel the smallest adequate section is 203×203×60 UC. For this section:

$P_f = 702$ kN $M_{fx} = 71.8$ kNm

Evaluating the combined effects gives:

$$\frac{513}{702} + \frac{12.8}{71.8} = 0.91 \leq 1.00 \quad \therefore \text{OK}$$

11.2.2 Concrete filled structural hollow sections

In the fire condition, the load on the column (as given in 11.2.1) is:

Compression load = 513 kN
Moment = 12.8 kNm

As explained above, the effective length of the column at the fire limit state

$$= 4.0 \times 0.7 = 2.8 \text{ m}$$

Using the safe load tables given in Design Data Sheet 10 (which have been calculated using the method in EC4-1-2 Annex G[5] and the method described in Reference [23]), select a 244.5×6.3 CHS with 4 reinforcing bars, 25 mm diameter, and grade 35 concrete and 50 mm axis distance.

For this section,
$EI_\theta = 2609 \text{ kNm}^2$, $\quad N_{fi,pl,Rd} = 2054 \text{ kN}$, $\quad M_{fx} = 60.4 \text{ kNm}$

For $l_\theta = 2.8$ m, the elastic critical load $N_{fi,cr}$ is given by:

$$N_{fi,cr} = \frac{\pi^2 EI_{fi,cr}}{l_\theta^2} = \frac{\pi^2 \, 2609.10^6}{2800^2} = 3284$$

Hence, the non-dimensional slenderness ratio, λ_θ is given by

$$\overline{\lambda}_\theta = \sqrt{N_{fi,pl,Rd}/N_{fi,cr}} = \sqrt{2054/3284} = 0.79$$

The buckling resistance is given by

$$N_{fi,Rd} = \chi \, N_{fi,pl,Rd}$$

Where

$$\chi = \frac{1}{\phi + [\phi^2 - \overline{\lambda}_\theta^2]^{0.5}}$$

and

$$\phi = 0.5 \, [1 + \alpha (\overline{\lambda}_\theta - 0.2) + \overline{\lambda}_\theta^2]$$

Hence, assuming $\alpha = 0.49$

$$\chi = 0.67$$

And thus

$$N_{fi,Rd} = 0.54 \times 2081 = 1376 \text{ kN}$$

Check if column is slender about x axis

Conservatively assume the ratio of moments in the x-direction equals unity. Hence $r_x = 1.0$

Section is slender if $\bar{\lambda}_\theta > 0.2(2 - r)$

$$0.2(2 - r_x) = 0.20$$

Column is slender, therefore

$$\beta_x = 0.66 + 0.44 r_x$$

$$k_x = \frac{\beta}{1 - \frac{N_f}{N_{fi,cr}}} = \frac{1.1}{1 - \frac{513}{3284}} = 1.3$$

There are no applied moments on the y-axis.

Check the combined effects

$$\frac{N_f}{N_{fi,Rd}} + k_x \frac{M_{fx}}{M_{fi,Rd}} + k_y \frac{M_{afy}}{M_{fi,Rd}} \leq 1.0$$

$$\frac{513}{1376} + 1.30 \frac{12.8}{60.4} + 0 = 0.65 \quad \therefore \text{OK}$$

11.3 Floor slab design

Design examples for the following types of floor are presented:

1. Shallow composite slabs with steel decking.

    These are used with conventional composite beams and with partially encased composite beams.

2. Deep composite slabs,

    These are used in slim floor construction.

11.3.1 Shallow composite slabs with steel decking

The structural layout of the floor is shown in Figure 11.7, in which the composite slab spans 3m. Design for the fire limit state can be carried out using the simplified method presented in SCI publication *The fire resistance of composite floors with steel decking*[20] (see Design Data Sheet 11), or alternatively using the fire engineering method, also presented in the same publication. BS 5850-8 permits the use of either method, and both methods are considered in the following example.

Simplified Method:

From Design Data Sheet 11:

As the imposed load is less than 6.7 kN/m^2, the minimum steel deck thickness is 0.9 mm and the minimum overall slab thickness is 120 mm. An A142 mesh with a top cover of between 15 and 45 mm is also required.

Fire Engineering Method:

This method is based on plastic moment resistance of the composite slab at the fire limit state, as permitted by BS 5950-8 and EC4-1-2. The difference between the codes is the calculation method used for obtaining the temperature distribution through the cross-section. The method of BS 5950-8, coupled with the material reduction factors in SCI publication[20], is explained below.

When using the fire engineering method, the slab thickness required to satisfy the insulation requirement of BS 476 (see Section 3.1) is given in Table F3 for trapezoidal decks and Table F4 for dovetail decks. In this example, a trapezoidal deck is being used together with lightweight concrete. From Table F3, the minimum thickness, measured above the top of the deck, is 60 mm. Hence, for practical reasons, a 110 mm overall slab depth is selected. This is 10 mm less than that used in the simplified method (above).

The total self weight of the lightweight composite slab, shown in Figure 11.9 is 1.83 kN/m^2

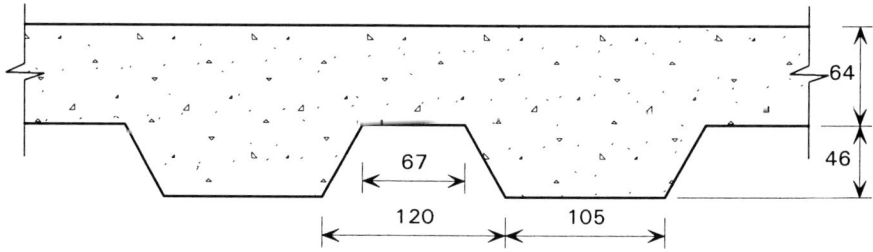

Figure 11.9 *Cross-section through the composite slab*

Table 11.6 *Design load summary for shallow floor slab*

Loading	Characteristic load (kN/m^2)	Partial load factors (BS 5950-8)	Design load in fire (kN/m^2)
Weight of composite floor slab	1.83	1.0	1.83
Raised floor, ceiling and services	0.7	1.0	0.7
Partitions	1.0	1.0	1.0
Imposed load	3.5	0.8	2.8
Total factored load at the fire limit state			**6.33**

Free design moment at the fire limit state

$M_0 = 6.33 \times 3.0^2 / 8 = 7.12$ kNm   per metre width.

Assume 8 mm bars in the ribs at 40 mm axis distance (Figure 11.10).

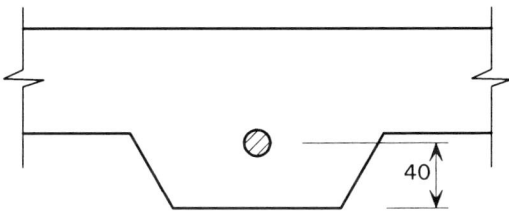

Figure 11.10   *Position of reinforcing bar*

Calculate sagging (positive) moment resistance:

Reinforcing bar temperature for 60 mins fire resistance = 290°C.
(See Table 12 in BS 5950-8 or Table 5 in SCI publication[20]).

Reduced tensile strength of the reinforcing bar:

$$p_r = \frac{k_r f_y}{\gamma_{mr}}$$

where:

k_r is the strength reduction factor for reinforcing bar = 1.0
(See Table 4 in SCI publication[20])

f_y is the reinforcement yield strength = 460 N/mm²

γ_m is the material partial safety factor for reinforcment = 1.0
(refer to BS 5950-8).

Therefore, $p_r = 460$ N/mm².

Tensile force in reinforcing bar,

$F_r = 460 \times \pi \times 4.0^2 = 23.1$ kN.

Strength of concrete in compression is given by;

$$p_c = \frac{0.67 f_{cu}}{\gamma_{mc}} k_r$$

where:

f_{cu} is the cube strength of concrete = 35 N/mm²

k_r is the strength reduction factor for concrete.
Assume concrete in compression is at a low enough temperature to use a value of 1.0 (see Table 4 in SCI publication[20]).

49

γ_m is the material safety factor for concrete = 1.3 (refer to BS 5950-8).

Therefore, p_c = 18.0 N/mm^2.

Depth of concrete stress block = (23.1 × 10^3) / (18.0 × 225) = 5.7 mm

∴ Lever arm = 110 - 40 - 5.7/2 = 67.1 mm

∴ Moment resistance = 23.1 × 67.1 × 10^{-3}
= 1.55 kNm per 225 mm width
= 6.9 kNm per metre width.

Calculate hogging (negative) moment of resistance:

Assume that the centre of the top reinforcing bar is 30 mm below the top surface of the floor. The lower exposed surface is in excess of 80 mm from the reinforcing bar, so that its full strength is assumed. (See Tables 4 and 5 in SCI publication[20].)

Therefore the tensile force in the reinforcing bar = 23.1 kN.

The concrete required to resist this force will be at varying temperatures, depending on the depth under consideration. The suggested procedure is to consider bands of concrete 10 mm thick. These layers are considered in sequence until sufficient compressive resistance is obtained. In Figure 11.11, two bands are shown. The outer 10 mm of concrete is normally ignored, because it contributes little compressive strength, due to high temperatures in this area.

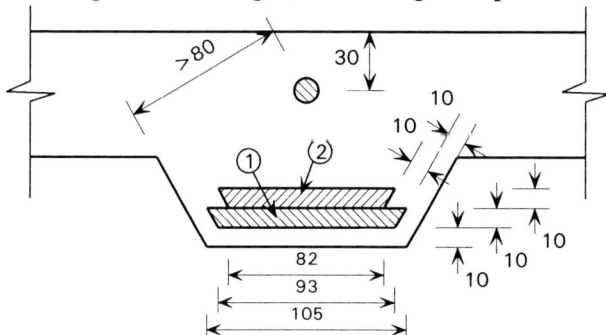

Figure 11.11 *Hogging moment resistance*

Consider band 1 (10 mm to 20 mm depth):

Average depth = 15 mm

Temperature = 550°C (refer to Table 12 in BS 5950-8 or Table 5 in SCI publication[20]).

k_r = 0.9 (see Table 4 in SCI publication[20]).

$$\therefore p_c = \frac{0.67}{\gamma_{mc}} f_{cu} \, k_r$$

Therefore, p_c = 16.2 N/mm^2

Area of band 1 = 10 (105 + 93)/2.0 = 990 mm^2

Compressive resistance of band 1 = 16.2 × 990 × 10^{-3} = 16.0 kN.

This is less than the tensile resistance of the reinforcement, so band 2 must be considered.

Consider band 2 (20 mm to 30 mm depth):

Average depth = 25 mm

Temperature = 430°C (refer to Table 12 in BS 5950-8 or Table 5 in SCI publication[20]).

k_r = 1.0 (refer Table 4 SCI publication[20]).

$$\therefore p_c = \frac{0.67}{\gamma_{mc}} f_{cu} k_r$$

Therefore p_c = 18.0 N/mm²

Area of band 2 = 10(93 + 82)/2.0 = 875 mm²

Compressive resistance of band 2 = 18.0 × 875 = 15.8 kN.

The combined resistance of band 1 and band 2 is greater than the tensile resistance of the reinforcement so only a portion of band 2 is required.

Area of band 2 required = (23.1 - 16.0) × 10^3 / 18.0 = 394 mm².

Approximate height in compression = 394/87.5 = 4.5 mm

Centroid of concrete stress block from bottom of slab
≈ 10 + (10+4.5)/2.0 = 17.3

Lever arm from centre of concrete to reinforcing bar = 110 - 30 - 17.3
= 62.7 mm

Moment resistance = 23.1 × 62.7 × 10-3
= 1.45 kNm per 225 mm width
= 6.45 kNm per metre width.

The following checks are carried out, as explained in SCI publication[20], according to the principles of plastic analysis

For an internal span, the condition for adequate moment resistance is given by:

$M_H + M_s \geq M_0$

where:

M_H is the hogging moment resistance in fire per unit width

M_S is the sagging moment resistance in fire per unit width

M_0 is the applied free moment per unit width.

∴ 6.45 + 6.9 = 13.35 > 7.12 ∴ OK

For an end span, the condition for adequate moment resistance is given by:

$$M_s + \frac{M_H}{2}\left(1 - \frac{M_H}{8 M_0}\right) \geq \text{OK}$$

$$6.9 + \frac{6.45}{2}\left(1 - \frac{6.45}{8\,(7.12)}\right) = 9.76 \geq 7.12 \therefore \text{OK}$$

Compared to the simplified method, the fire engineering method permits the use of a thinner slab, although additional reinforcement is required.

11.3.2 Deep composite slabs

The composite slab spans 6.0 m between beams for the structural layout shown in Figure 11.2, and for slim floor construction. The PMF SD225 deck used is unpropped during construction. The example uses the SD225 deck with an overall composite slab depth of 300 mm (Figure 12.12). Design in fire is carried out to EC4-1-2. Similar to the previous example, this analysis consists of calculating the plastic moment of resistance at a given fire resistance period (in this case 60 minutes).

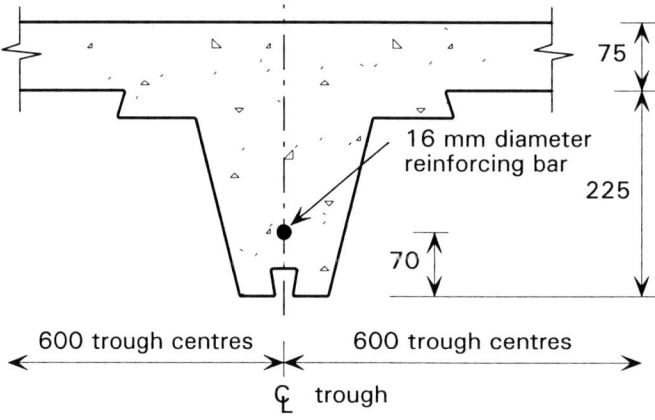

Figure 11.12 *Section through deep composite slab*

For lightweight concrete, total self weight of composite slab = 2.76 kN/m²

Table 11.7 *Design load summary for deep floor slab*

Loading	Characteristic load (kN/m²)	Partial load factors (BS 5950-8)	Design load in fire (kN/m²)
Raised floor, ceiling and services	0.7	1.0	0.7
Partitions	1.0	1.0	1.0
Weight of composite floor slab	2.76	1.0	2.76
Imposed load	3.5	0.8	2.8
Total factored load at the fire limit state			**7.26**

Free design moment at the fire limit state

$M_o = 7.26 \times 6.0^2 / 8.0 = 32.7$ kNm.

EC4-1-2 permits a simple method for calculating reinforcement temperatures. However, recent fire tests have shown that the EC4 temperatures are unconservative for this type of steel deck (see Section 6.2). This is due to the method in EC4 being based on fire tests on shallow decks. Therefore, for SD225 decking, recommend reinforcement temperatures and axes distances are taken from Table 6.1 of this publication.

Assume concrete grade 30/35, cylinder strength, $f_{c,20°}$ = 30 N/mm²

Reinforcement (16 mm diameter bar), $f_{ys,20°C}$ = 460 N/mm²

Reinforcement temperature (Table 6.1) = 398 °C

From Table 3.4 EC4-1-2, $f_{smax,\phi} / f_{sy,20°C}$ = 0.94

Therefore tensile strength of reinforcing bar = 0.94 × 460 = 432 N/mm²

Tensile resistance of reinforcing bar = 432 × 201 = 86912 N

Assume the depth of concrete in compression = y
Compressive resistance of concrete = 600 y × 0.85 × 30 = 15300 y N
(Refer to cl. 4.3.1.4 (1) EC4-1-2 for 0.85 factor)

To balance tensile resistance of reinforcement, y = 5.7 mm

Lever arm = 300 - 7 - 5.7/2 = 227 mm

Moment of resistance = 227×86912
= 19.7 kNm per 600 mm width
= 32.9 kNm per metre width.

Since applied free moment is less than the moment of resistance, the deep deck slab achieves 60 minutes fire resistance.

12 REFERENCES

1. THE BUILDING REGULATIONS
   (a) Approved Document B, Fire Safety
   Department of the Environment and The Welsh Office
   The Stationery Office, 1991

   (b) The Building Standards (Scotland) Regulations
   Scottish Office
   The Stationery Office, 1990

   (c) The Building Regulations (Northern Ireland)
   Department of the Environment
   The Stationery Office, 1990

2. BRITISH STANDARDS INSTITUTION
   BS 5950: Structural use of steelwork in building
   BS 5950-8:1990, Code of practice for fire resistant design

3. BRITISH STANDARDS INSTITUTION
   BS 476: Fire tests on building materials and structures
   BS 476-20:1987: Method of determination of the fire resistance of elements of construction (general principles)
   BS 476-21: 1987: Method of determination of the fire resistance of load bearing elements of construction

4. BRITISH STANDARDS INSTITUTION
   ENV 1993: Eurocode 3: Design of steel structures
   ENV 1993 -1-2:1995: Structural fire design
   (DD ENV 1993-1-2, including UK NAD, *in preparation*)

5. BRITISH STANDARDS INSTITUTION
   ENV 1994: Eurocode 4: Design of composite steel and concrete structures
   ENV 1994-1-2:1994: Structural fire design
   (DD ENV 1994-1-2, including UK NAD, *in preparation*)

6. BRITISH STANDARDS INSTITUTION
   ENV 1991: Eurocode 1: Basis of design and actions on structures
   DD ENV 1991-1: 1996: Basis of design (including UK NAD)
   ENV1991-2-2: 1996: Actions on structures exposed to fire

7. BRITISH STANDARDS INSTITUTION
   BS 5950: The Structural Use of Steelwork in Buildings
   BS 5950-1:1990, Code of Practice for Design in Simple and Continuous Construction

8. MULLETT, D.L.
   Slim floor design and construction (SCI-P-110)
   The Steel Construction Institute, 1992

9. MULLETT, D.L. and LAWSON, R.M.
Design of *Slimflor* fabricated beams using deep decking (SCI-P-248)
The Steel Construction Institute, 1999

10. LAWSON R.M. MULLETT, D.L. and RACKHAM, J.W.
Design of asymmetric *Slimflor* beams using deep composite decking (SCI-P-175)
The Steel Construction Institute, 1997

11. MULLETT, D.L.
Design of RHS *Slimflor* edge beams (SCI-P-169)
The Steel Construction Institute, 1997

12. NEWMAN, G.M.
Technical Report: The fire resistance of shelf angle floor beams to BS 5850: Part 8 (SCI-P-126)
The Steel Construction Institution, 1993

13. BRITISH STANDARDS INSTITUTION
ENV 1994: Eurocode 4: Design of composite steel and concrete structures
DD ENV 1994-1-1:1994: General rules and rules for buildings (including UK NAD)

14. THE STEEL CONSTRUCTION INSTITUTE and THE BRITISH CONSTRUCTIONAL STEELWORK ASSOCIATION
Joints in simple construction - Volume 1: Design methods (Second Edition)
SCI & BCSA, 1993

15. THE STEEL CONSTRUCTION INSTITUTE and THE BRITISH CONSTRUCTIONAL STEELWORK ASSOCIATION
Joints in steel construction: Moment connections
SCI & BCSA, 1995

16. ASSOCIATION FOR SPECIALIST FIRE PROTECTION
Fire protection for structural steel in buildings (2nd Edition - Revised)
ASFP/SCI/FTSG, 1992

17. BUILDING RESEARCH ESTABLISHMENT
Fire resistant steel structures: Free standing blockwork filled columns and stanchions
BRE Digest 317, 1986

18. NEWMAN, G.M.
Technical Report: The fire resistance of web-infilled steel columns (SCI-P-124)
The Steel Construction Institute, 1992

19. MORRIS, W.A. READ, R.E.H. and COOKE, G.M.E.
    Guidelines for the construction of fire-resisting structural elements
    Building Research Establishment, 1982

20. NEWMAN, G.M.
    The fire resistance of composite floors with steel decking (2nd edition) (SCI-P-056)
    The Steel Construction Institute, 1991

21. BRITISH STANDARDS INSTITUTION
    BS 8110: Structural use of concrete
    BS 8110-1:1997, Code of practice for design and construction

22. LAW, M and O'BRIEN, T.
    Fire safety of bare external structural steel (SCI-P-009)
    The Steel Construction Institute, 1981

23. NEWMAN, G.M, SIMMS, W.I
    Technical Report: The fire resistance of concrete filled structural hollow sections to Eurocode 4
    The Steel Construction Institute, (Expected 1999)

APPENDIX A: BS 5950-8: Limiting temperature method

The method uses "load ratio" as a non-dimensional measure of the load resisted by a member. Load ratio is defined as:

$$\text{Load ratio} = \frac{\text{Applied load or moment at time of fire}}{\text{Load or moment resistance at } 20\,^\circ\text{C}}$$

After the load ratio is calculated, the limiting temperatures (or maximum allowable temperatures) can be obtained from Table 5 in BS 5950-8. This table is reproduced below (Table A.1).

For unprotected steel to be used, the limiting temperatures are compared against the design temperatures which are tabled in BS 5950-8 and correspond to the expected maximum temperature of the section for a given period of fire resistance (i.e, 30 and 60 minutes). These design temperatures depend on the flange thickness of the section.

If the limiting temperature is less than the design temperature, specified in BS 5950-8, then the beam does not require fire protection. Realistically, only 15 or 30 minutes fire resistance can be achieved.

Table A.1 *Limiting temperatures and load ratios from BS 5950-8*

Description of member	Limiting temperature(°C) at a load ratio of:					
	0.7	0.6	0.5	0.4	0.3	0.2
Members in compression						
$\lambda \leq 70$	510	540	580	615	655	710
$\lambda > 70$ but ≤ 180	460	510	545	590	635	635
Members in bending supporting concrete or composite slabs						
a) Unprotected beams or beams protected with ductile protection	590	620	650	680	725	780
b) Other protected beams	540	585	625	655	700	745
Members in bending not supporting concrete slabs						
a) Unprotected beams or beams protected with ductile protection	520	555	585	620	660	715
b) Other protected beams	460	510	545	590	635	690
Members in tension	460	510	545	590	635	690

APPENDIX B: The moment capacity method

The moment capacity method is based on plastic design. For bare beams it is only applicable for sections which satisfy the requirements for a plastic or compact section, as defined in BS 5950-1[7].

The method forms the basis of the calculation for shelf angle beams, slim floor beams and partially encased beams. The calculation procedure is summarised below:

1. The temperature distribution through the beam at the required fire resistance time period is obtained. This is either computed or obtained from tests.

2. The cross-section is divided into rectangular elements of approximately equal temperature. For each element reduced strength is calculated at elevated temperatures using the strength reduction factors given in Table 1 in BS 5950-8[2].

3. The 'plastic' neutral axis of the heated section is determined, such that in the absence of axial forces, the net tensile and compressive forces acting on the section are in equilibrium. The moment capacity (or moment resistance) is then calculated.

   Note that often the cross section may be considered to act compositely with any concrete encasement at the fire limit state, and therefore the compressive resistance of the concrete may be included in the analysis of the cross section depending on the temperature distribution in the concrete.

4. The applied moment is calculated using the fire limit state load factors.

5. If the moment capacity (resistance) is greater than the applied moment then the section is adequate and protection is not required.

An example of the division into elements of a shelf angle beam is shown in Figure B1. The division between elements 4 and 5 corresponds to a steel temperature of 350°C; below this temperature, the steel may be assumed to be at full strength.

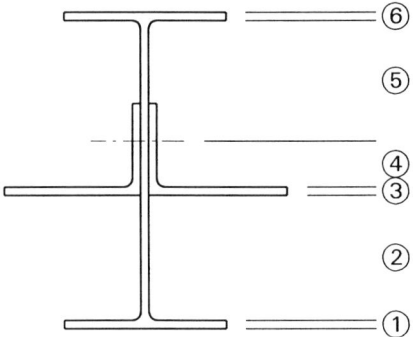

Figure B.1 *Division of a shelf angle beam into elements*

APPENDIX C: Design of 'open' car parks in fire conditions.

Approved Document B (England and Wales), Technical Booklet E (N. Ireland) and Technical Standards Part D (Scotland) all specify 15 minutes fire resistance for most open sided car parks and periods of fire resistance ranging from 30 to 120 minutes for all other cases. The Approved document states that a car park is considered open sided provided it complies with the following requirements:

1. There are no basement storeys.

2. Each storey is be naturally ventilated by permanent openings at each car parking level, having an aggregate ventilation area not less than 5% of the floor area at that level, of which at least half should be in two opposing walls.

3. If the building is used for any other purpose, the part forming the car park should be a separated compartment.

Provided the load ratio is less than or equal to 0.6 (see Section 3.33) then all Universal Beam Sections will achieve 15 minutes fire resistance, except those sections shown in Table C.1. For these sections, 15 minutes fire resistance may be achieved if the load ratio does not exceed the value in Table C.1.

Table C.1 *Maximum load ratios to achieve 15 minutes fire resistance*

UB section size	Maximum load ratio to achieve 15 minutes fire resistance
127 × 76 × 13	0.54
152 × 89 × 16	0.54
178 × 102 × 19	0.54
203 × 133 × 25	0.57
254 × 102 × 22	0.49
254 × 102 × 25	0.56
305 × 102 × 25	0.48
305 × 102 × 28	0.58
356 × 127 × 33	0.58
406 × 140 × 39	0.59

All Universal Column Sections, except for 152×152×23, used as columns, will achieve 15 minutes fire resistance, provided the load ratio is less than or equal to 0.6.

APPENDIX D: Design Data Sheets

Design Data Sheet No.	Structural Member	Page No.
1	Slimfor beams	63
2	Rectangular hollow section edge beam	64
3	Asymmetrical slim floor beams	65
4	Shelf angle floor beam	68
5	Partially encased beams	71
6	Unreinforced concrete web-infilled columns	72
7	Connection strength for slim floor beams	73
8	Connection strength for ASB slim floor beams	74
9	Reinforced concrete web-infilled columns	76
10	Concrete infilled hollow section columns	78
11	Composite slabs with profiled metal deck	82

DESIGN DATA SHEET 1
Slimflor Beams

60 Minutes Fire Resistance

SFB fabricated from: Universal Column & plate. The plate is 200 mm wider than the UC flange.
Steel grade: S355.
Steel decking: SD255.
Concrete: Normal weight (NWC) or lightweight (LWC) grade 30.
Slab depth: Minimum depth is 295mm for NWC and 285mm for LWC. For composite design, where only one value is given it applies to both concrete types. In non-composite design, deeper beams may project above the slab.
(For full design information see *Design of Slimflor fabricated beams using deep composite decking*[9])

Universal Column	Non-composite design				Composite design (NWC and LWC)		
	Slab depth (mm)		Moment resistance (kNm)		Slab depth (mm)	Minimum effective width of slab (mm)	Moment resistance (kNm)
	NWC	LWC	NWC	LWC			
152x152x30	295	285	136	129	295 (285)	1840	204 (193)
152x152x37	295	285	151	143	295 (285)	1840	242 (228)
203x203x46	295	285	200	190	295 (288)	1840	237 (222)
203x203x52	295	285	217	207	295 (291)	1840	250 (242)
203x203x60	295	285	242	232	295	1840	271
203x203x71	295	285	272	263	301	1880	308
203x203x86	295	285	328	318	307	1920	372
254x254x73	295	285	290	280	339	2119	386
254x254x89	295	285	346	336	345	2158	454
254x254x107	295	285	425	417	352	2198	550
254x254x132	295	285	554	546	361	2258	700
254x254x167	295	285	760	755	374	2338	940

Universal Column	Non-composite design				Composite design (NWC and LWC)		
	Slab depth (mm)		Moment resistance (kNm)		Slab depth (mm)	Minimum effective width of slab (mm)	Moment resistance (kNm)
	Full*	Mod*	Full*	Mod*			
305x305x97	308	295	423	417	395	2456	571
305x305x118	315	295	521	512	400	2497	681
305x305x137	321	295	633	619	406	2534	804
305x305x158	327	295	770	749	412	2576	955
305x305x198	340	295	1062	1022	425	2656	1275
305x305x240	353	295	1403	1331	438	2734	1650
305x305x283	365	295	1740	1537	450	2814	2028
356x368x129	356	295	634	628	441	2754	805
356x368x153	362	295	786	776	447	2794	971
356x368x177	368	295	958	941	453	2833	1158
356x368x202	375	295	1150	1122	460	2873	1366

Notes:

In the upper part of the table, for non-composite construction, the slab depth is greater than the depth of the Universal Column section so the appropriate minimum slab depth is 295mm for NWC and 285mm for LWC for 60 minutes fire resistance.

In the lower part of the table, for non-composite construction, Full* indicates that the slab depth is equal to the depth of the Universal Column section (values apply to both NWC and LWC) and Mod* indicates that the slab depth is less than the depth of the Universal Column section (i.e. modified construction). For NWC the appropriate slab depth is 295mm (as shown); for LWC the slab depth may be reduced to 285mm, without affecting the moment resistance and load ratio.

For composite construction, in all cases, the slab depth is 85 mm greater than the depth of the Universal Column section. The effective width of slab is based on one quarter of the beam span for a sensible range of uses.

DESIGN DATA SHEET 2
Asymmetric *Slimflor* Beams

60 Minutes Fire Resistance			
Steel grade: S355.			
Steel decking: SD255 or CF210			
Concrete: Normal weight (NWC) or lightweight (LWC) grade 30.			
Reference: For full design information see *Design of asymmetric Slimflor beams using deep composite decking*[10]			
Section	Span (mm)	Effective width (mm)	Moment resistance (kNm)
280 ASB 100 (using normal weight concrete, with minimum 35 mm cover over top flange or lightweight concrete with 30mm cover to top flange)	5500	688	269/265+
	6000	750	271/266
	6500	813	273/268
	7000	875	275/270
	7500	938	277/271
	8000	1000	279/273
280 ASB 136 (using lightweight or normal weight concrete, with minimum 30mm cover to top flange)	5500	688	384
	6000	750	387
	6500	813	389
	7000	875	391
	7500	938	394
	8000	1000	396
300 ASB 153 (using lightweight or normal weight concrete, with minimum 30mm cover to top flange)	5500	688	478
	6000	750	481
	6500	813	484
	7000	875	487
	7500	938	490
	8000	1000	492
300 ASB 153 (using lightweight concrete with slab flush with top flange)	5500	688	445*
	6000	750	447*
	6500	813	449*
	7000	875	451*
	7500	938	453*
	8000	1000	455*

\+ For 280 ASB 100 the two values of moment resistance are for normal weight and lightweight concrete.

* These values differ from those in SCI publication[10] as a result of a re-appraisal of the design method.

DESIGN DATA SHEET 3
Rectangular Hollow Section Edge Beam

60 Minutes Fire Resistance — Non-composite Edge Beam					
Edge Beam fabricated from: RHS & plate (15mm). The plate is 120 mm wider than the RHS section.					
Steel grade: S355.					
Steel decking: SD255.					
Concrete: Normal weight (NWC) or lightweight (LWC) grade 30.					
Slab thickness: Minimum and applicable to SD225 steel deck					
Note: Tabulated values are for decking perpendicular to edge beam. For the case where the deck runs parallel to the edge beam, the moment resistance in fire should be *reduced by 5%*					
Reference: For full design information see *Design of RHS Slimflor Edge Beams*[11]					
Size of RHS section	Slab depth (mm)		Moment resistance (kNm)		Load ratio
	NWC	LWC	Ultimate limit state	Fire limit state	
200 × 100 × 5	295	285	108.1	47.2	0.44
200 × 100 × 6.3	295	285	137.4	57.7	0.42
200 × 100 × 8	295	285	167.2	71.6	0.43
200 × 100 × 10	295	285	200.6	89.2	0.44
200 × 100 × 12.5	295	285	235.7	112.1	0.48
200 × 100 × 16	295	285	275.9	142.8	0.52
200 × 150 × 5	295	285	130.1	60.3	0.46
200 × 150 × 6.3	295	285	161.1	73.4	0.46
200 × 150 × 8	295	285	197.3	90.6	0.46
200 × 150 × 10	295	285	240.2	112.5	0.47
200 × 150 × 12.5	295	285	285.9	141.4	0.49
200 × 150 × 16	295	285	341.4	181.6	0.53
250 × 150 × 5	295	285	179.5	86.4	0.48
250 × 150 × 6.3	295	285	223.3	105.9	0.47
250 × 150 × 8	295	285	282.1	131.1	0.46
250 × 150 × 10	295	285	337.0	162.8	0.48
250 × 150 × 12.5	295	285	398.7	204.0	0.51
250 × 150 × 16	295	285	473.5	262.1	0.55
300 × 200 × 6.3	330	330	333.0	167.3	0.50
300 × 200 × 8	330	330	416.2	207.4	0.50
300 × 200 × 10	330	330	508.8	257.6	0.51
300 × 200 × 12.5	330	330	605.6	323.4	0.53
300 × 200 × 16	330	330	724.0	416.6	0.58
300 × 250 × 6.3	330	330	369.0	191.1	0.52
300 × 250 × 8	330	330	466.4	236.3	0.51
300 × 250 × 10	330	330	567.7	293.4	0.52
300 × 250 × 12.5	330	330	683.5	368.3	0.54
300 × 250 × 16	330	330	825.2	476.0	0.58

Table continues....

Size of RHS section	Slab depth (mm)		Moment resistance (kNm)		Load ratio
	NWC	LWC	Ultimate limit state	Fire limit state	
200 × 120 × 5	295	285	120.1	52.4	0.44
200 × 120 × 6	295	285	141.5	61.3	0.43
200 × 120 × 6.3	295	285	145.4	63.9	0.44
200 × 120 × 8	295	285	180.9	79.4	0.44
200 × 120 × 10	295	285	216.8	98.6	0.45
200 × 120 × 12.5	295	285	256.1	123.9	0.48
250 × 100 × 6.3	295	285	195.0	86.0	0.44
250 × 100 × 8	295	285	239.9	106.9	0.45
250 × 100 × 10	295	285	284.0	133.0	0.47
250 × 100 × 12.5	295	285	330.4	166.1	0.50
250 × 100 × 16	295	285	386.8	212.6	0.55
260 × 140 × 6.3	295	290	235.6	108.9	0.46
260 × 140 × 8	295	290	289.6	135.1	0.47
260 × 140 × 10	295	290	346.7	167.8	0.48
260 × 140 × 12.5	295	290	409.0	210.0	0.51
260 × 140 × 16	295	290	484.3	269.6	0.56
300 × 100 × 6.3	330	330	264.7	118.9	0.45
300 × 100 × 8	330	330	320.2	148.6	0.46
300 × 100 × 10	330	330	376.0	184.9	0.49
300 × 100 × 12.5	330	330	436.6	231.2	0.53
300 × 100 × 16	330	330	512.3	296.6	0.58
200 × 200 × 5	295	285	149.1	73.3	0.49
200 × 200 × 6.3	295	285	187.1	88.6	0.47
200 × 200 × 8	295	285	227.3	109.0	0.48
200 × 200 × 10	295	285	278.0	135.3	0.49
200 × 200 × 12.5	295	285	335.4	170.0	0.51
200 × 200 × 16	295	285	403.7	219.4	0.54
250 × 250 × 6.3	295	285	286.2	145.0	0.51
250 × 250 × 8	295	285	357.6	178.4	0.50
250 × 250 × 10	295	285	432.6	221.2	0.51
250 × 250 × 12.5	295	285	524.5	277.7	0.53
250 × 250 × 16	295	285	636.5	359.0	0.56
300 × 300 × 6.3	330	330	404.4	214.6	0.53
300 × 300 × 8	330	330	510.5	264.5	0.52
300 × 300 × 10	330	330	621.8	328.6	0.53
300 × 300 × 12.5	330	330	757.3	412.5	0.54
300 × 300 × 16	330	330	923.3	534.8	0.58

60 Minutes Fire Resistance — Composite Edge Beam

Edge Beam fabricated from: RHS & plate (15mm). The plate is 120 mm wider than the RHS section.
Steel grade: S355.
Steel decking: SD255.
Concrete: Normal weight (NWC) or lightweight (LWC) grade 30.
Slab thickness: Minimum and applicable to SD225 steel deck
Notes:
1. The table is based on 40% shear connection and can conservatively be used for higher degrees of shear connection.
2. Tabulated values are for decking perpendicular to edge beam. For the case where the deck runs parallel to the edge beam, the moment resistance in fire should be *reduced by 5%*

Reference: For full design information see *Design of RHS Slimflor Edge Beams*[11]

Section size (mm)	Normal weight concrete				Lightweight concrete			
	Slab depth (mm)	Moment resistance (kNm)		Load ratio	Slab depth (mm)	Moment resistance (kNm)		Load ratio
		Ultimate limit state	Fire limit state			Ultimate limit state	Fire limit state	
200 × 100 × 8	295	265.5	136.4	0.51	285	248.0	124.4	0.50
200 × 100 × 10	295	293.2	157.1	0.54	285	276.4	144.5	0.52
200 × 100 × 12.5	295	322.6	182.6	0.57	285	306.3	169.4	0.55
200 × 100 × 16	295	358.7	216.5	0.60	285	342.0	202.8	0.59
200 × 150 × 8	295	301.5	153.5	0.51	285	285.5	142.0	0.50
200 × 150 × 10	295	339.0	178.5	0.53	285	321.0	166.5	0.52
200 × 150 × 12.5	295	380.7	210.4	0.55	285	362.9	197.9	0.55
200 × 150 × 16	295	429.6	254.6	0.59	285	412.6	240.5	0.58
250 × 150 × 8	335	453.0	228.7	0.50	335	453.0	228.7	0.50
250 × 150 × 10	335	499.1	267.7	0.54	335	499.1	267.7	0.54
250 × 150 × 12.5	335	550.8	315.9	0.57	335	550.8	315.9	0.57
250 × 150 × 16	335	616.3	381.3	0.62	335	616.3	381.3	0.62
200 × 120 × 8	295	280.7	143.2	0.51	285	262.6	131.7	0.50
200 × 120 × 10	295	311.4	165.8	0.53	285	295.0	153.2	0.52
200 × 120 × 12.5	295	346.5	194.2	0.56	285	329.4	181.0	0.55
250 × 100 × 8	335	400.9	209.6	0.52	335	400.9	209.6	0.52
250 × 100 × 10	335	435.5	242.8	0.56	335	435.5	242.8	0.56
250 × 100 × 12.5	335	474.7	283.0	0.60	335	474.7	283.0	0.60
250 × 100 × 16	335	525.7	336.5	0.64	335	525.7	336.5	0.64
260 × 140 × 8	345	480.8	244.6	0.51	345	480.8	244.6	0.51
260 × 140 × 10	345	525.3	285.8	0.54	345	525.3	285.8	0.54
260 × 140 × 12.5	345	578.0	335.7	0.58	345	578.0	335.7	0.58
260 × 140 × 16	345	644.3	404.5	0.63	345	644.3	404.5	0.63
200 × 200 × 8	295	330.7	170.0	0.51	285	312.8	159.1	0.51
200 × 200 × 10	295	379.6	199.8	0.53	285	356.9	187.9	0.53
200 × 200 × 12.5	295	433.6	237.4	0.55	285	415.8	225.6	0.54
200 × 200 × 16	295	497.5	290.8	0.58	285	480.2	277.5	0.58
250 × 250 × 8	335	537.7	266.4	0.50	335	537.7	266.4	0.50
250 × 250 × 10	335	609.2	316.3	0.52	335	609.2	316.3	0.52
250 × 250 × 12.5	335	690.3	379.1	0.55	335	690.3	379.1	0.55
250 × 250 × 16	335	791.4	469.1	0.59	335	791.4	469.1	0.59

DESIGN DATA SHEET 4
Shelf Angle Floor Beams

The design procedure for shelf angle beams is described in Appendix E of BS 5950-8. When choosing a shelf angle configuration, the following points should be noted:

- To ensure sufficient insulation to the top of the beam the precast slab should not have any deliberately designed voids in the end 75 mm. Also the void between the precast slab and beam should be filled with grout.

- The angles should not be less than $125 \times 75 \times 12$ (S355) with the longer leg horizontal and the vertical leg pointing upwards. The angles are considered to contribute to the strength of the section in fire.

- The precast slab should have a minium bearing of 75 mm on the angles.

- The transverse moment capacity of the angles needs to be checked at the required fire resistance period.

The calculation procedure is summarised below.

(1) Determine the temperature distribution across the section. BS 5950-8[2] Appendix E provides tabulated data which enables the temperature distribution to be obtained.

(2) The moment capacity method (Appendix B, of this publication) is used to calculate the moment resistance of the section.

(3) If the moment of resistance is greater than the applied moment no fire protection is required.

(4) Calculate the transverse moment capacity of the angles using the strength reduction factors for 1.5% strain. This must be greater than the applied transverse moment due to the loads transmitted via the slab at the fire limit state.

For normal, design the contribution of the angles to the bending resistance of the beam is ignored. However, in fire the angles are assumed to act structurally with the beam. Therefore, as well as checking vertical capacity, the bolt or weld connection of the angle to the beam should be checked to ensure adequate resistance to the horizontal shear force necessary to develop the required axial forces in the angles.

Safe load tables for transverse bending of the angles, and tables of longitudinal forces between the beam and angles, are presented in SCI publication *The fire resistance of shelf angle floor beams to BS 5950: Part 8*[12].

30 Minutes Fire Resistance - Shelf Angle Floor Beams

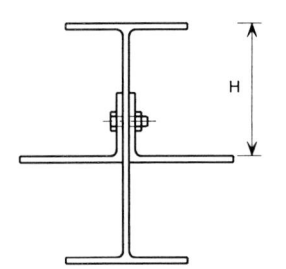

Shelf angle: 125 × 75 × 12, or larger
Steel grade: Angles, S355; UB Section, S275 or S355
Notes: The dimension H gives the highest position of the upper surface of horizontal leg of the angle for the given load ratio to be achieved.

The moment resistance in fire = M_r × load ratio

M_r is the ultimate moment capacity (cold) of the Universal Beam which forms the shelf angle beam.

Design information is based on BS 5950-8.

Section size	Steel Grade S275				Steel Grade S355			
	M_r (kNm)	H for a Load ratio of:			M_r (kNm)	H for a Load ratio of:		
		0.4	0.5	0.6		0.4	0.5	0.6
305 × 102 × 25	93	82	82	108	120	82	93	125
305 × 102 × 28	112	84	84	118	145	84	100	133
305 × 102 × 33	132	86	90	125	170	86	105	139
305 × 127 × 37	149	86	94	124	192	86	108	137
305 × 127 × 42	168	87	98	128	217	87	112	140
305 × 127 × 48	194	89	104	133	251	89	116	145
305 × 165 × 40	172	85	97	128	222	85	110	141
305 × 165 × 46	199	87	100	131	257	87	112	143
305 × 165 × 54	232	89	104	136	300	89	116	148
356 × 127 × 33	148	84	113	149	192	86	127	160
356 × 127 × 39	180	86	119	157	232	89	132	165
356 × 171 × 45	213	86	123	162	275	96	136	170
356 × 171 × 51	246	88	126	167	318	98	138	173
356 × 171 × 57	277	90	127	169	358	101	139	175
356 × 171 × 67	333	96	134	174	430	104	145	182
406 × 140 × 39	198	98	145	185	256	112	158	193
406 × 140 × 46	244	99	149	189	315	113	162	199
406 × 178 × 54	288	110	152	196	372	122	163	207
406 × 178 × 60	328	109	150	197	424	121	162	209
406 × 178 × 67	370	111	154	201	478	122	165	213
406 × 178 × 74	414	115	159	207	534	126	170	215
457 × 152 × 52	301	124	178	221	388	137	190	229
457 × 152 × 60	353	121	179	225	456	133	190	231
457 × 152 × 67	396	124	183	229	512	136	194	235
457 × 152 × 74	430	128	187	232	560	139	198	240
457 × 152 × 82	477	125	187	235	621	136	198	242
457 × 191 × 67	405	130	176	227	522	142	187	238
457 × 191 × 74	456	130	178	230	588	142	189	242
457 × 191 × 82	504	134	182	235	651	145	193	247
457 × 191 × 89	534	135	184	238	695	146	195	250
457 × 191 × 98	591	129	180	237	770	140	191	249
533 × 210 × 82	565	161	215	272	730	172	224	283
533 × 210 × 92	651	163	219	278	840	173	228	290
533 × 210 × 101	694	164	222	281	904	175	231	295
533 × 210 × 109	748	161	223	282	974	172	229	297
533 × 210 × 122	849	154	227	282	1105	165	227	299

60 Minutes Fire Resistance - Shelf Angle Floor Beam

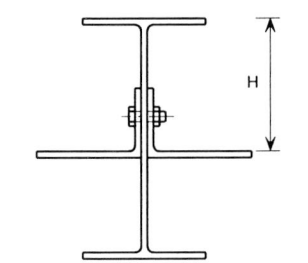

Shelf angle: 125 × 75 × 12, or larger S355 steel
Steel grade: Angle, S355; UB Section, S275 or S355
Notes: The dimension H gives the highest position of the upper surface of horizontal leg of the angle for the given load ratio to be achieved.

The moment resistance in fire = M_r × load ratio

M_r is the ultimate moment capacity (cold) of the Universal Beam which forms the shelf angle beam.

Design information is based on BS 5950-8.

Section size	Steel grade S275				Steel Grade S355			
	M_r (kNm)	H for a Load ratio of:			M_r (kNm)	H for a Load ratio of:		
		0.4	0.5	0.6		0.4	0.5	0.6
305 × 102 × 25	93	119	149	175	120	129	158	184
305 × 102 × 28	112	127	157	184	145	137	167	193
305 × 102 × 33	132	135	165	192	170	144	174	201
305 × 127 × 37	149	136	165	192	192	145	174	202
305 × 127 × 42	168	141	170	197	217	150	181	207
305 × 127 × 48	194	148	177	206	251	158	190	212
305 × 165 × 40	172	139	169	192	222	149	174	204
305 × 165 × 46	199	145	175	199	257	159	183	208
305 × 165 × 54	232	155	179	210	300	171	196	218
356 × 127 × 33	148	149	182	212	192	158	191	221
356 × 127 × 39	180	157	191	221	232	166	201	234
356 × 171 × 45	213	162	196	222	275	173	211	236
356 × 171 × 51	246	169	205	231	318	184	218	249
356 × 171 × 57	277	176	215	241	358	194	223	255
356 × 171 × 67	333	191	226	260	430	211	241	267
406 × 140 × 39	198	175	212	245	256	185	223	259
406 × 140 × 46	244	184	223	259	315	196	239	278
406 × 178 × 54	288	191	232	261	372	207	251	281
406 × 178 × 60	328	201	244	274	424	219	264	296
406 × 178 × 67	370	211	255	285	478	230	266	302
406 × 178 × 74	414	220	267	297	534	241	276	307
457 × 152 × 52	301	209	253	294	388	224	271	315
457 × 152 × 60	353	219	267	310	456	238	287	332
457 × 152 × 67	396	228	277	322	512	247	298	336
457 × 152 × 74	430	236	286	332	560	257	308	343
457 × 152 × 82	477	245	296	342	621	266	319	355
457 × 191 × 67	405	229	278	311	522	248	300	335
457 × 191 × 74	456	241	291	325	588	261	305	343
457 × 191 × 82	504	250	302	337	651	271	312	347
457 × 191 × 89	534	256	310	345	695	279	320	356
457 × 191 × 98	591	267	313	352	770	290	332	369
533 × 210 × 82	565	278	336	376	730	299	360	402
533 × 210 × 92	651	295	355	396	840	317	366	408
533 × 210 × 101	694	303	364	406	904	326	375	417
533 × 210 × 109	748	310	371	413	974	335	384	427
533 × 210 × 122	849	325	375	417	1105	349	399	444

DESIGN DATA SHEET 5
Partially Encased Beams

Preliminary Design Data										
Notes: The table gives minimum cross-sectional dimensions and reinforcement of a partially encased composite beam, based on EC4-1-2 Total reinforcement is expressed as a ratio of the cross sectional area of one flange.										
Load level (η)	Minimum cross sectional dimensions (mm) and minimum reinforcement									
	R30		R60		R90		R120		R180	
	min b	reinf	min b	reinf	min b	reinf	min b	reinf	min b	reinf
$\eta = 0.3$										
$h \geq 0.9$ min b	70	0.0	100	0.0	170	0.0	200	0.0	260	0.0
$h \geq 1.5$ min b	60	0.0	100	0.0	150	0.0	180	0.0	240	0.0
$h \geq 2$ min b	60	0.0	100	0.0	150	0.0	180	0.0	240	0.0
$\eta = 0.5$										
$h \geq 0.9$ min b	80	0.0	170	0.0	250	0.4	270	0.5	-	-
$h \geq 1.5$ min b	80	0.0	150	0.0	200	0.2	240	0.3	300	0.5
$h \geq 2$ min b	70	0.0	120	0.0	180	0.2	220	0.3	280	0.3
$h \geq 3$ min b	60	0.0	100	0.0	170	0.2	200	0.3	250	0.3
$\eta = 0.7$										
$h \geq 0.9$ min b	80	0.0	270	0.4	300	0.6	-	-	-	-
$h \geq 1.5$ min b	80	0.0	240	0.3	270	0.4	300	0.6	-	-
$h \geq 2$ min b	70	0.0	190	0.3	210	0.4	270	0.5	320	1.0
$h \geq 3$ min b	70	0.0	170	0.2	190	0.4	270	0.5	300	0.8

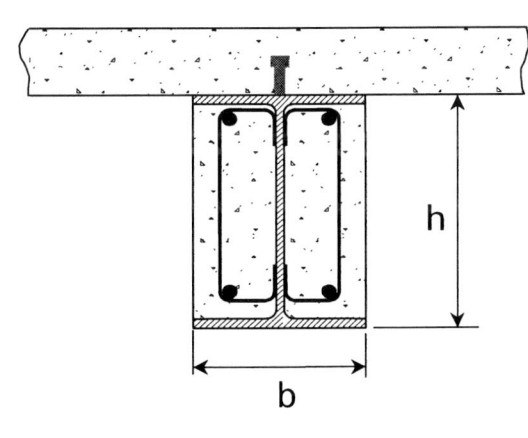

DESIGN DATA SHEET 6
Connection Strength of Slim Floor Beams

The Table gives the "cold" moment capacity for a range of full depth end plates and bolt sizes. The "cold" value, rather than a value at elevated temperatures, is tabulated because the connection will be shielded by the deep composite floor slab and will remain relatively cool in fire.

60 Minutes Fire Resistance	
Steel grade:	*Slimflor* beam, UC and plate grade S355 connected to columns grade S275
Connection detail:	Full depth (flush) welded end plates (grade S275), welds 8 mm fillet
Notes:	Floor slab assumed to cover connection.
	Connection assumed to be non-composite.

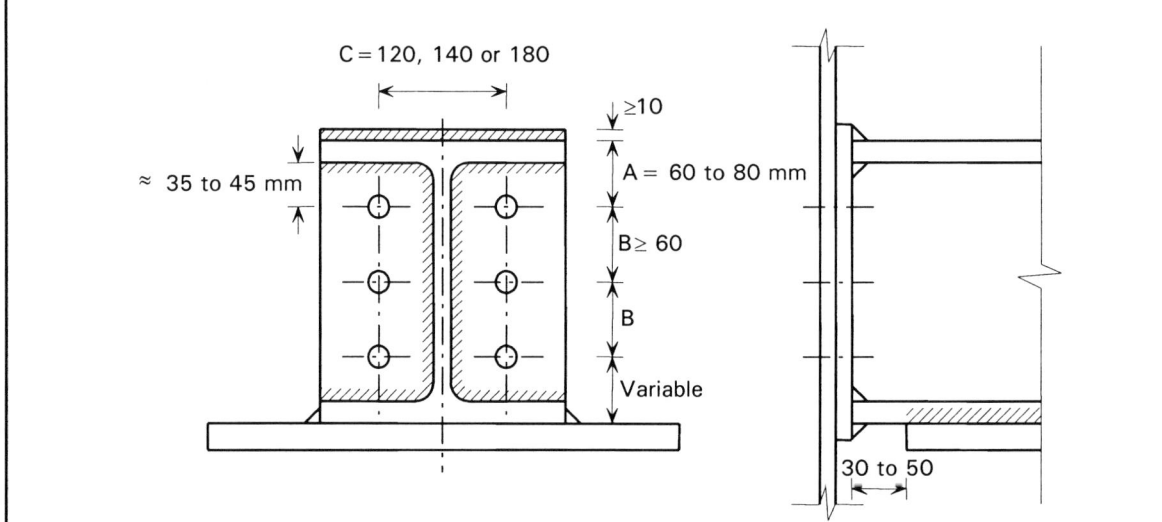

Slimflor UC size	Column size (all weights)	Moment capacity (kNm)	
		10mm plate, M20 bolts	12mm plate, M24 bolts
203UC	203UC	21	30(23)*
(all weights)	254UC	21	30
4 bolts	305UC	21	30
C = 120 mm	356UC	21	30
254UC	203UC	28	38(29)*
(all weights)	254UC	32	45
6 bolts	305UC	32	45
C = 140 mm	356UC	32	45
305UC	203UC	Not practical	Not practical
(all weights)	254UC	35#	50#
6 bolts	305UC	40#	58#
C = 180 mm	356UC	40#	58#

* denotes lower value for 203x203x46 UC
#Not recommended for *Slimflor* constructed using Universal Columns with weights greater than 200 kg/m

DESIGN DATA SHEET 7

Connection Strength for Asymmetric *Slimflor* Beams

The Table gives the "cold" moment capacity for a range of full depth end plates and bolt sizes. The "cold" value, rather than a value at elevated temperatures, is tabulated because the connection will be shielded by the deep composite floor slab and will remain relatively cool in fire.

60 Minutes Fire Resistance	
Steel grade:	ASB grade S355 connected to columns grade S275
Connection detail:	Full depth (flush) welded end plates (grade S275), welds 8mm fillet
Notes:	Floor slab assumed to cover connection.
	Connection assumed to be non-composite.

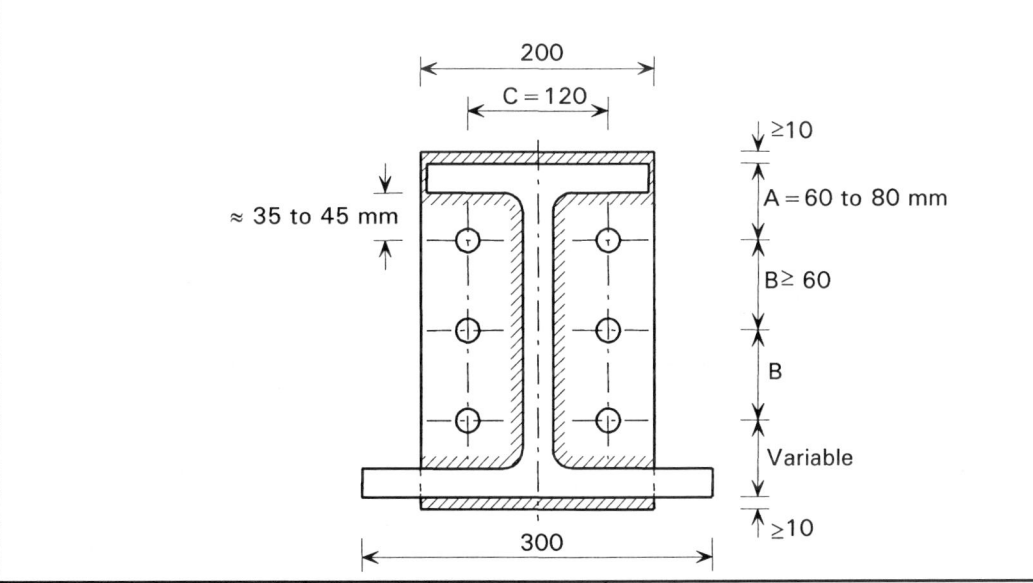

ASB size	Column size(S275) (all weights)	Moment capacity (kNm)	
		10mm plate, M20 bolts	12mm plate, M24 bolts
280 ASB (all weights) 6bolts	203UC	34	47(36)*
	254UC	34	49
	305UC	34	49
	356UC	34	49
300 ASB (all weights) 6bolts	203UC	39	52(40)*
	254UC	39	56
	305UC	39	56
	356UC	39	56
300 ASB (all weights) 8bolts	203UC	46	61(47)*
	254UC	46	66
	305UC	46	66
	356UC	46	66
* denotes lower value for 203x203x46			

DESIGN DATA SHEET 8
Unreinforced Concrete Web-infilled Columns

Two separate tables are given for S275 and S355 steel for 60 minutes fire resistance. In each table the compressive resistance (kN) and moment resistance are given. The compressive resistance is tabulated for a range of system lengths (system length is equal to storey height). In fire, the effective length is assumed to be 0.7 of the system length (see 5.1). Full design information may be obtained from SCI publication *The fire resistance of web-infilled columns* [18].

The combined effects of axial load and moment are evaluated using the following linear interaction formula:

$$\frac{P}{P_f} + \frac{M_y}{M_{fy}} + \frac{M_x}{M_{fx}} = 1.0$$

Where:

P is the axial load
M_x is the applied moment about x-x axis
M_y is the applied moment about y-y axis
P_f is the compressive resistance
M_{fx} is the bending resistance about x-x axis
M_{fy} is the bending resistance about y-y axis

60 Minutes Fire Resistance - S275 Steel								

Notes: Minimum concrete strength: 33 N/mm² (cylinder), 40 N/mm² (cube)
Compressive resistance is calculated assuming that the column is continuous at both ends.
For columns which are continuous at one end multiply the resistance by 0.82.

Section size	Moment resistance (kNm)		Compressive resistance, P_f (kN) for column system lengths L (mm)				
	M_{fx}	M_{fy}	2500	3000	3500	4000	4500
203 × 203 × 46	40.3	22.9	744	682	616	551	488
203 × 203 × 52	47.0	25.4	803	738	669	600	533
203 × 203 × 60	56.7	29.0	882	811	735	659	586
203 × 203 × 71	69.9	34.6	974	899	820	740	662
203 × 203 × 86	92.2	43.6	1153	1064	971	876	784
254 × 254 × 73	100.2	56.0	1499	1415	1326	1233	1137
254 × 254 × 89	126.6	66.5	1671	1579	1482	1381	1276
254 × 254 × 107	163.7	80.9	1914	1812	1703	1590	1473
254 × 254 × 132	219.7	104.5	2258	2139	2015	1884	1748
254 × 254 × 167	307.1	14.3	2800	2656	2505	2347	2182
305 × 305 × 97	183.8	105.1	2411	2307	2198	2084	1964
305 × 305 × 118	229.0	122.6	2655	2544	2428	2307	2180
305 × 305 × 137	281.4	142.7	2967	2843	2714	2578	2437
305 × 305 × 158	342.2	167.0	3282	3147	3006	2859	2706
305 × 305 × 198	475.8	220.4	3941	3782	3616	3443	3262
305 × 305 × 240	636.2	284.6	4785	4595	4398	4191	3976
305 × 305 × 283	792.4	346.3	5570	5320	5067	4810	4548

60 Minutes Fire Resistance - S355 Steel

Notes: Minimum concrete strength: 33 N/mm² (cylinder), 40 N/mm² (cube)
Compressive resistance is calculated assuming that the column is continuous at both ends.
For columns which are continuous at one end multiply the resistance by 0.82.

Section size	Moment resistance (kNm)		Compressive resistance, P_f (kN) for column system lengths L (mm)				
	M_{fx}	M_{fy}	2500	3000	3500	4000	4500
203 × 203 × 46	50.9	26.0	833	750	663	578	501
203 × 203 × 52	59.4	29.2	912	824	732	641	557
203 × 203 × 60	71.8	33.6	998	902	801	702	610
203 × 203 × 71	89.6	41.1	1130	1027	919	810	709
203 × 203 × 86	118.2	52.7	1336	1214	1086	957	837
254 × 254 × 73	126.7	63.6	1670	1564	1449	1326	1202
254 × 254 × 89	161.7	77.2	1894	1776	1648	1512	1373
254 × 254 × 107	209.4	95.7	2199	2065	1921	1766	1608
254 × 254 × 132	281.8	126.1	2629	2474	2305	2126	1940
254 × 254 × 167	394.9	173.4	3320	3129	2922	2700	2471
305 × 305 × 97	232.5	118.5	2671	2544	2408	2261	2105
305 × 305 × 118	292.4	141.2	2982	2845	2698	2541	2374
305 × 305 × 137	359.9	167.0	3367	3212	3047	2869	2681
305 × 305 × 158	438.4	198.2	3765	3596	3415	3220	3014
305 × 305 × 198	611.1	266.9	4655	4450	4231	3995	3745
305 × 305 × 240	818.8	349.7	5653	5409	5148	4868	4571
305 × 305 × 283	1045	437.8	6644	6322	5988	5642	5285

DESIGN DATA SHEET 9
Reinforced Concrete Web-infilled Columns

30, 60, 90 and 120 Minutes Fire Resistance

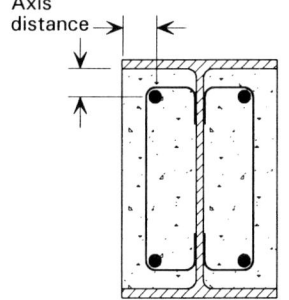

Notes: This has been developed using the design method in Annex F of EC4-1-2 and gives section sizes together with sizes and number of reinforcing bars required to achieve the required fire resistance period.

The minimum required axis distance is defined as the distance from an external surface to the centre of the bar.

The following assumptions were used in deriving the table:

1. The system length (system length is equal to storey height) is 15 times the width of the column.
2. In fire an effective length factor of 0.7 is used.
3. When calculating the load level the reinforcement is ignored in the 'cold' design.

load level (η)	Section size and reinforcement details for columns with fire resistance of:			
	R30	R60	R90	R120
0.3	All sections (Minimum reinforcement to control spalling)	All sections 203 × 203 × 46 and larger with 4 f16 (Minimum axis distance = 40 mm)	356 × 406 × 393 + 4 f40 356 × 406 × 340 + 4 f40 356 × 406 × 287 + 4 f32 356 × 406 × 235 + 4 f32 356 × 368 × 202 + 4 f25 256 × 268 × 177 + 4 f25 356 × 368 × 153 + 4 f20 356 × 368 × 129 + 4 f16 305 × 305 × 198 + 4 f32 305 × 305 × 137 + 4 f25 305 × 305 × 188 + 4 f25 305 × 305 × 97 + 4 f20 254 × 254 × 132 + 4 f32 254 × 254 × 107 + 4 f32 254 × 254 × 89 + 4 f32 254 × 254 × 73 + 4 f25 (Minimum axis distance = 50 mm)	356 × 368 × 202 + 4 f40 356 × 368 × 177 + 4 f40 356 × 368 × 153 + 4 f40 356 × 368 × 129 + 4 f32 (Minimum axis distance = 60 mm)
0.5	All sections 203 × 203 × 46 and larger with 4 f12 except : 203 × 203 × 86 + 4 f16 (Minimum axis distance = 30 mm)	356 × 406 × 287 + 4 f40 356 × 406 × 235 + 4 f32 356 × 368 × 202 + 4 f32 356 × 368 × 177 + 4 f32 356 × 368 × 153 + 4 f25 356 × 368 × 129 + 4 f25 305 × 305 × 137 + 4 f32 305 × 305 × 118 + 4 f25 305 × 305 × 97 + 4 f25 254 × 254 × 132 + 4 f32 254 × 254 × 107 + 4 f32 254 × 254 × 89 + 4 f25 254 × 254 × 73 + 4 f25 (Minimum axis distance = 40 mm)	356 × 368 × 129 + 4 f40 305 × 305 × 97 + 4 f40 (Minimum axis distance = 50 mm)	No sections meet design criteria

load level (η)	Section size and reinforcement details for columns with fire resistance of:			
	R30	R60	R90	R120
0.7	356 × 406 × 634 +4 f40 356 × 406 × 551 +4 f40 356 × 406 × 467 +4 f40 356 × 406 × 393 +4 f32 356 × 406 × 340 +4 f32 356 × 406 × 287 +4 f25 356 × 406 × 235 +4 f25 356 × 368 × 202 +4 f20 356 × 368 × 177 +4 f20 356 × 368 × 153 +4 f16 356 × 368 × 129 +4 f12 305 × 305 × 283 +4 f32 305 × 305 × 240 +4 f32 305 × 305 × 198 +4 f25 305 × 305 × 137 +4 f20 305 × 305 × 118 +4 f20 305 × 305 × 97 +4 f16 254 × 254 × 167 +4 f32 254 × 254 × 132 +4 f25 254 × 254 × 107 +4 f25 254 × 254 × 89 +4 f20 254 × 254 × 73 +4 f20 (Minimum axis distance = 30 mm)	356 × 356 × 129 +4 f40 (Minimum axis distance = 40 mm)	No sections meet design criteria	No sections meet design criteria

DESIGN DATA SHEET 10
Concrete Filled Hollow Section Columns

Two tables, one for circular hollow sections and the other for square hollow sections are given. Each Table includes design information for 30 and 60 minutes fire resistance. The design method is based on EC4-1-2 Annex G.

The moment and buckling resistances, from the tables, may be combined using the following linear interaction formula[23].

$$\frac{N_f}{N_{fi,Rd}} + k_x \frac{M_{fx}}{M_{fi,Rd}} + k_y \frac{M_{afy}}{M_{fi,Rd}} \leq 1.0$$

where:

	N_f	is the applied axial load
	$N_{fi,Rd}$	is the buckling resistance
	$N_{cr,f}$	is the Euler buckling load
	M_{fx}	is the applied moment about x-x axis
	$M_{fi,Rd}$	is the moment capacity
	M_{afy}	is the accidental moment about y-y axis
	k_x, k_y	is the moment magnification factors

Generally, for a practical column in a non sway building, k_x and k_y are both equal to unity, and for preliminary design purposes may be assumed to be unity. However, for final design, the values should always be determined. The following procedures may be used, for bending about each axis

(1) Check whether the column is 'non-slender'.

A column is deemed 'non-slender' for any axis if the slenderness on that axis, $\bar{\lambda}_\theta$ is not greater than $\bar{\lambda}_{crit}$

where

$$\bar{\lambda}_\theta = \sqrt{\frac{N_{fi,pl,Rd}}{N_{cr,f}}}$$

$$N_{cr,f} = \frac{\pi EI}{l^2}$$

$$\bar{\lambda}_{crit} = 0.2(2 - r)$$

r is the ratio of the smaller to the larger end moment, i.e. $-1 \leq r \leq +1$
(in cases were transverse loading occurs along the column, r may be taken as 1.0.)

The design tables below give the effective "EI" for the section in fire, for use in the evaluation of $N_{cr,f}$.

If the column is non-slender, k may be taken as 1.0.

(2) If k is not 1.0 from (1) above, check whether second order effects are significant.

Second order effects are significant when:

$$\frac{N_f}{N_{cr,f}} > 0.1$$

If second order effects are not significant, k may be taken as 1.0.

(3) If k is not 1.0 from (1) or (2) above, calculate the value of k from the following:

$$k = \frac{\beta}{1 - \frac{N_f}{N_{cr,f}}} , \text{ but not less than } 1.0$$

where, $\beta = 0.66 + 0.44\,r$ but ≥ 0.44

For columns with transverse loading within the column length, β should be taken as 1.0

The following table gives guidance on the magnitude of k when it is calculated according the expression in (3) above and gives values for $\bar{\lambda}_{crit}$ when checking (1) above.

$\frac{N_f}{N_{cr,f}}$	Ratio of the smaller to the larger end moment, r							
	≤0.3	-0.2	0	0.2	0.4	0.6	0.8	1
0.1	1	1	1	1	1	1.03	1.12	1.22
0.2	1	1	1	1	1.05	1.16	1.27	1.38
0.3	1	1	1	1.07	1.19	1.32	1.45	1.57
0.4	1	1	1.1	1.25	1.39	1.54	1.69	1.83
0.5	1.06	1.14	1.32	1.5	1.67	1.85	2.02	2.2
0.6	1.32	1.43	1.65	1.87	2.09	2.31	2.53	2.75
0.7	1.76	1.91	2.2	2.49	2.79	3.08	3.37	3.67
0.8	2.64	2.86	3.3	3.74	4.18	4.62	5.06	5.5
0.9	5.28	5.72	6.6	7.48	8.36	9.24	10.1	11
$\bar{\lambda}_{crit}$	0.46	0.44	0.4	0.36	0.32	0.28	0.24	0.2

Users should note that the concrete strength is expressed as cylinder strength, as is normal in Eurocodes.

Concrete Filled Square Hollow Section

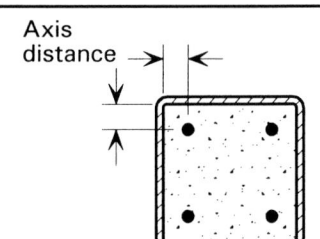

Notes: Resistances are calculated assuming S275 steel and are conservative for S355 steel.

Concrete grade is the cylinder strength.

Axis distance = distance from the inside of the steel section to the centre of the reinforcing bar.

Make-up of concrete filled hollow section				Resistances		Stiffness in fire	Buckling Resistance (kN)		
Section Size	Concr. Grade	Bar dia. (mm)	Axis dist. (mm)	Bending $M_{fi,Rd}$ (kNm)	Axial $N_{fi,pl,Rd}$ (kN)	EI (kNm²)	Effective Lengths (mm)		
							2000	3000	4000
30 Minutes									
160 × 160 × 5.0	25	16	30	29.5	998	1476	828	670	512
160 × 160 × 5.0	35	16	30	30.0	1178	1511	951	748	555
180 × 180 × 5.0	25	20	30	48.7	1374	2603	1187	1001	803
180 × 180 × 5.0	35	20	30	49.6	1611	2668	1363	1125	880
200 × 200 × 5.0	25	20	40	54.2	1559	3575	1382	1196	993
200 × 200 × 5.0	35	20	40	55.0	1860	3691	1617	1373	1111
250 × 250 × 6.3	25	25	40	116.0	2498	10300	2354	2147	1922
250 × 250 × 6.3	35	25	40	118.4	2980	10650	2772	2504	2208
300 × 300 × 6.3	25	25	40	160.5	3175	20787	3096	2894	2681
300 × 300 × 6.3	35	25	40	162.7	3892	21665	3754	3483	3194
350 × 350 × 8	25	25	40	237.1	4128	41210	4125	3914	3698
350 × 350 × 8	35	25	40	239.9	5111	43034	5060	4775	4480
350 × 350 × 10	25	25	40	274.4	4357	45935	4357	4149	3928
350 × 350 × 10	35	25	40	278.2	5321	47730	5287	4999	4703
400 × 400 × 10	25	25	40	352.5	5327	76062	5327	5184	4955
400 × 400 × 10	35	25	40	356.6	6611	79447	6611	6357	6045
60 Minutes									
160 × 160 × 5.0	25	16	30	10.7	500	459	373	271	188
160 × 160 × 5.0	35	16	30	10.9	620	468	437	300	202
180 × 180 × 5.0	25	20	30	19.3	747	899	595	462	338
180 × 180 × 5.0	35	20	30	19.8	921	919	702	521	368
200 × 200 × 5.0	25	20	40	38.0	1195	1647	978	781	589
200 × 200 × 5.0	35	20	40	38.5	1426	1686	1133	875	639
250 × 250 × 6.3	25	25	40	84.8	2000	5305	1805	1587	1347
250 × 250 × 6.3	35	25	40	85.9	2402	5453	2126	1838	1523
300 × 300 × 6.3	25	25	40	112.7	2565	11052	2426	2219	1994
300 × 300 × 6.3	35	25	40	114.0	3185	11488	2966	2680	2366
350 × 350 × 8.0	25	25	40	152.6	3300	21634	3219	3008	2787
350 × 350 × 8.0	35	25	40	154.3	4172	22622	4016	3721	3406
350 × 350 × 10	25	25	40	160.0	3326	22497	3250	3042	2823
350 × 350 × 10	35	25	40	162.6	4181	23448	4035	3744	3435
400 × 400 × 10	25	25	40	199.4	4140	38938	4124	3906	3682
400 × 400 × 10	35	25	40	201.5	5301	40889	5222	4912	4591

Concrete Filled Circular Hollow Section

Notes: Resistances are calculated assuming S275 steel and are conservative for S355 steel.

Concrete grade is the cylinder strength.

Axis distance = distance from the inside of the steel section to the centre of the reinforcing bar.

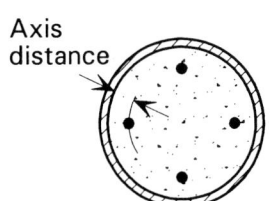

Axis distance

Make-up of concrete filled hollow section				Resistances		Stiffness in fire	Buckling Resistance (kN)		
Section Size	Concr. Grade	Bar dia. (mm)	Axis dist. (mm)	Bending $M_{fi,Rd}$ (kNm)	Axial $N_{fi,pl,Rd}$ (kN)	EI (kNm²)	Effective Lengths (mm)		
							2000	3000	4000
30 Minutes									
168.3 × 5.0	25	16	40	22.8	953	1061	747	569	410
168.3 × 5.0	35	16	40	24.6	1118	1091	848	626	440
193.7 × 5.0	25	20	40	37.9	1342	1986	1114	901	689
193.7 × 5.0	35	20	40	41.0	1568	2052	1271	1004	748
219.1 × 5.0	25	20	40	49.5	1545	3206	1352	1155	942
219.1 × 5.0	35	20	40	53.1	1841	3327	1579	1323	1052
244.5 × 6.3	25	25	50	78.3	2167	5675	1952	1715	1453
244.5 × 6.3	35	25	50	84.7	2536	5886	2252	1951	1622
273 × 6.3	25	25	50	99.3	2455	8711	2282	2060	1815
273 × 6.3	35	25	50	106.2	2922	9075	2682	2394	2077
323.9 × 6.3	25	32	50	178.7	3616	18372	3465	3199	2915
323.9 × 6.3	35	32	50	192.6	4289	19194	4068	3730	3363
355.6 × 8.0	25	32	50	234.3	4193	28749	4101	3840	3567
355.6 × 8.0	35	32	50	249.1	4999	29992	4847	4513	4159
406.4 × 10.0	25	32	50	340.2	5222	52702	5221	4955	4683
406.4 × 10.0	35	32	50	357.4	6272	54942	6224	5880	5527
457 × 10.0	25	32	50	425.1	6102	81306	6102	5911	5639
457 × 10.0	35	32	50	444.1	7447	85117	7447	7135	6774
60 Minutes									
168.3 × 5.0	25	16	40	13.8	673	383	427	272	175
168.3 × 5.0	35	16	40	15.1	781	390	468	287	183
193.7 × 5.0	25	20	40	26.4	1051	839	754	527	358
193.7 × 5.0	35	20	40	28.7	1218	857	838	566	376
219.1 × 5.0	25	20	40	34.9	1226	1415	968	744	541
219.1 × 5.0	35	20	40	38.2	1459	1456	1113	826	583
244.5 × 6.3	25	25	50	55.1	1752	2529	1448	1166	887
244.5 × 6.3	35	25	50	60.4	2054	2609	1655	1299	963
273 × 6.3	25	25	50	70.4	1992	4129	1742	1488	1214
273 × 6.3	35	25	50	76.8	2384	4291	2043	1710	1359
323.9 × 6.3	25	32	50	137.1	3059	9926	2819	2526	2204
323.9 × 6.3	35	32	50	150.0	3638	10336	3308	2929	2511
355.6 × 8.0	25	32	50	168.8	3437	14916	3253	2977	2677
355.6 × 8.0	35	32	50	182.0	4143	15562	3872	3509	3112
406.4 × 10.0	25	32	50	225.6	4133	26740	3306	3033	2738
406.4 × 10.0	35	32	50	238.6	5073	27991	3919	3563	3173
457 × 10.0	25	32	50	278.0	4871	43452	4028	3762	3483
457 × 10.0	35	32	50	291.2	6091	45784	4890	4535	4156

DESIGN DATA SHEET 11
Composite Slabs with Profiled Metal Deck

1. Simplified Method

The following design advice is taken from SCI publication *Fire resistance of composite floors with steel decking* [20].

The following recommendations apply:

- The slab must be continuous over at least one support.

- The loads on the floor, excluding self weight, should not exceed 6.7 kN/m^2. For increased load see reference 20.

- A142, A193 or A252 mesh reinforcement, satisfying the ductility requirements of BS4449: 1988, should be provided. The mesh size depends on the span and fire resistance. The reinforcement should have a top cover of between 15 and 45 mm.

- The tabulated data is for spans up to 3.6 m. Longer spans are possible; see reference 20 for these cases.

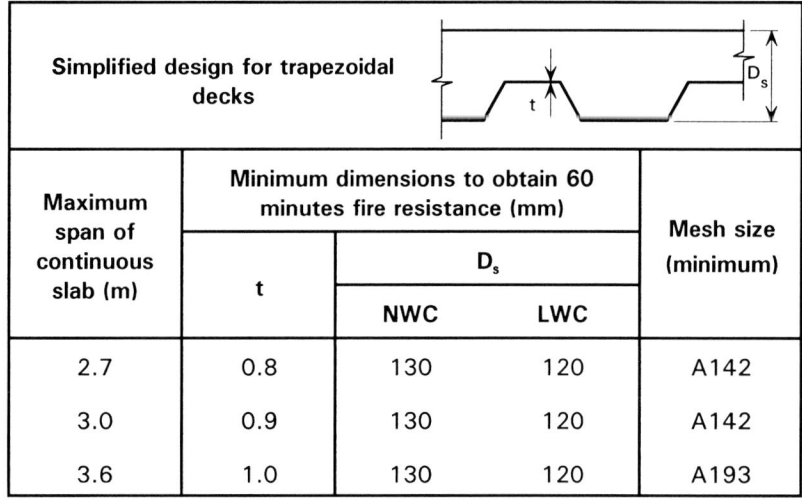

Simplified design for trapezoidal decks			
Maximum span of continuous slab (m)	Minimum dimensions to obtain 60 minutes fire resistance (mm)		Mesh size (minimum)
	t	D_s	
		NWC / LWC	
2.7	0.8	130 / 120	A142
3.0	0.9	130 / 120	A142
3.6	1.0	130 / 120	A193

NWC = Normal weight concrete   LWC = Lightweight concrete.

The design data is valid for deck profiles of 45 to 60 mm depth. For deck profiles of depth D less than 55 mm and spans not greater than 3 m, the slab depth may be reduced by (55 - D) up to a maximum reduction of 10 mm. For deck profiles greater than 60 mm the slab depth should be increased by (D - 60). Dovetail details that protrude above the nominal top of the deck profile can normally be ignored provided they are not greater than 10 mm in height.

Simplified design for dovetail decks				
Maximum span of continuous slab (m)	Minimum dimensions to obtain 60 minutes fire resistance (mm)			Mesh size (minimum)
	t	D_s		
		NWC	LWC	
2.5	0.8	100	100	A142
3.0	0.9	120	110	A142
3.6	1.0	125	120	A193

NWC = Normal weight concrete   LWC = Lightweight concrete

The design data is valid for deck profiles of 38 to 50 mm depth. For deck profiles depth D greater than 50 mm the slab depth should be increased by (D-50).

2. Fire Engineering Method

The following Tables give the minimum slab thicknesses for use with the "fire engineering method". The Tables are taken from BS 5950-8 and SCI publication[20].

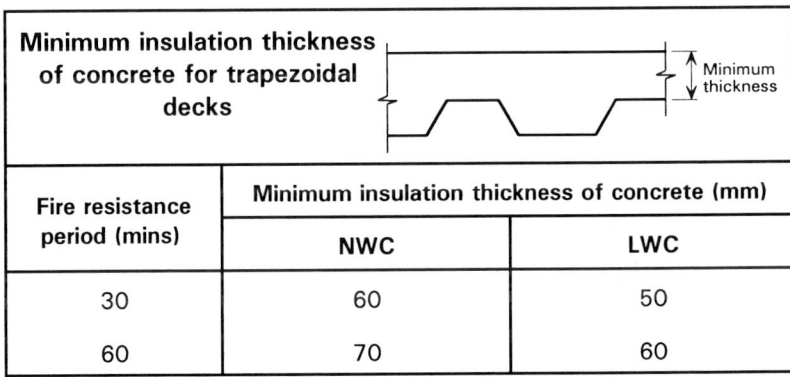

Minimum insulation thickness of concrete for trapezoidal decks		
Fire resistance period (mins)	Minimum insulation thickness of concrete (mm)	
	NWC	LWC
30	60	50
60	70	60

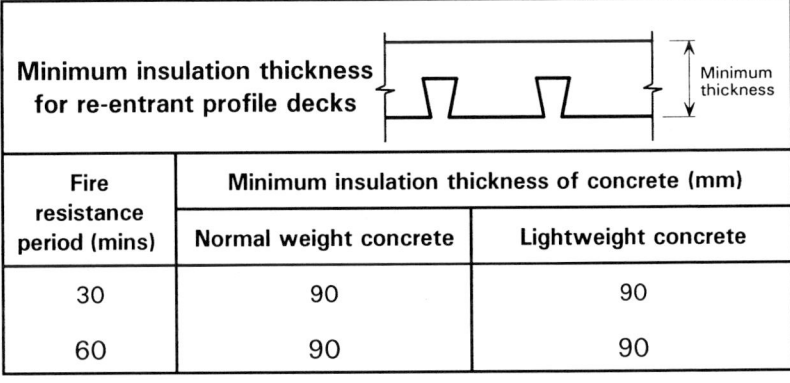

Minimum insulation thickness for re-entrant profile decks		
Fire resistance period (mins)	Minimum insulation thickness of concrete (mm)	
	Normal weight concrete	Lightweight concrete
30	90	90
60	90	90

Typeset and page make-up by The Steel Construction Institute, Ascot, Berks. SL5 7QN
Printed and bound by Alden Press, Osney Mead, Oxford, OX2 0EF
1000 - 7/99, 300 - 3/01 (BCF710)